SpringerBriefs in Applied Sciences and Technology

SpringerBriefs present concise summaries of cutting-edge research and practical applications across a wide spectrum of fields. Featuring compact volumes of 50 to 125 pages, the series covers a range of content from professional to academic.

Typical publications can be:

- A timely report of state-of-the art methods
- An introduction to or a manual for the application of mathematical or computer techniques
- A bridge between new research results, as published in journal articles
- A snapshot of a hot or emerging topic
- An in-depth case study
- A presentation of core concepts that students must understand in order to make independent contributions

SpringerBriefs are characterized by fast, global electronic dissemination, standard publishing contracts, standardized manuscript preparation and formatting guidelines, and expedited production schedules.

On the one hand, **SpringerBriefs in Applied Sciences and Technology** are devoted to the publication of fundamentals and applications within the different classical engineering disciplines as well as in interdisciplinary fields that recently emerged between these areas. On the other hand, as the boundary separating fundamental research and applied technology is more and more dissolving, this series is particularly open to trans-disciplinary topics between fundamental science and engineering.

Indexed by EI-Compendex, SCOPUS and Springerlink.

Juraj Ružbarský

Contactless System for Measurement and Evaluation of Machined Surfaces

Springer

Juraj Ružbarský
Department of Technical Systems Design and Monitoring
Faculty of Manufacturing Technologies
Technical University of Košice
Košice, Slovakia

ISSN 2191-530X ISSN 2191-5318 (electronic)
SpringerBriefs in Applied Sciences and Technology
ISBN 978-3-031-08980-0 ISBN 978-3-031-08981-7 (eBook)
https://doi.org/10.1007/978-3-031-08981-7

This Springer imprint is published by the registered company Springer Nature Switzerland AG
The registered company address is: Gewerbestrasse 11, 6330 Cham, Switzerland

Preface

The work deals with the research and development of the contactless system for the measurement and evaluation of selected characteristics of machined surfaces. Under existing structural, software, and hardware solutions available in the market it describes the design and construction of new equipment designed for contactless characterization of surface geometry working based on laser profilometry by triangulation principle. Further, the work focuses on the quality assessment of surfaces produced by the technology of abrasive waterjet cutting (AWJ) and by laser. The work analyzes the performed experiment of measurement of machined surfaces of samples of aluminum, stainless steel, and constructional steel using the contactless method with the LPM system and contact method using Mitutoyo SJ 400 surface roughness tester. Nine samples were produced. During the manufacturing process based on the technology of abrasive waterjet cutting the samples differed in shift speed of a cutting head which resulted in different quality of final surfaces. The evaluated parameters were Ra—average arithmetic deviation of assessed profile and Rz—maximum height of profile roughness. Consequently, the values measured by the contactless method were compared with the values having been measured by the contact method, i.e. with the Mitutoyo SJ 400 roughness tester.

Košice, Slovakia Juraj Ružbarský

Acknowledgment This document is part of the project and was supported by the project KEGA 015TUKE-4/2022.

Introduction

Contrary to the ideal surface, fine micro-geometric surface irregularities are referred to as surface roughness. Development of science, advancement of technical branches, and increase of product quality using progressive materials require both precise technical and economical prescription of manufacturing demands by design engineers as well as a precisely determined check of products to observe defined surface roughness. The product surface structure considerably influences overall product quality. It affects service life, reliability, strength, fatigue, appearance, and slip properties. Electric resistance, heat transfer, magnetic properties, losses caused by friction, wear, and corrosion resistance and consequent noisiness, running-in period, and corrosion fatigue rank among other physical and chemical properties induced by the preciseness of manufacturing of functional areas of machine components.

Research, comparison, application of knowledge on surface roughness, and consequent evaluation of surface roughness all result in the quality increase and higher precision of component functionality and effectiveness as well as reliability of entire equipment. Especially reliability should be considered as the determining factor by engineers. The reasonable design of a design engineer can contribute to the selection of correct manufacturing technology and conditions of the manufacturing process. Surface quality improvement of machine components decreases overall component weight dimensioned for particular loading conditions. High-quality machined surface simplifies further processing, engineering or finishing operations, and surface treatment which either ecologically or economically affects the manufacturing of the respective product. Last but not least it should be mentioned that appropriate requirements related to surface roughness can considerably influence the final price. Therefore roughness should be selected as gross as possible yet it still should conform to the correct area function. Surface quality refers to surface roughness as well as to the uncertainty of dimensions, geometrical shape and position, and component surface hardness. The surface issue is dealt with by topography which in general focuses on the description of the examined object. 3D topographic methods then refer to such measurement methods the result of which is the so-called topographic deflection z (x, y). The deflection is defined as the distance of a point (x, y) from the reference plane. The reference plane is realized either physically during method calibration or it is specified by the used method principle. These topographic methods can be divided into two categories—the contact methods and the optical methods.

Contents

Symbols and Abbreviations

μm	Micrometre
Ra	Arithmetical mean deviation of the profile (μm)
Rz	Height of the profile roughness (μm)
Rp	Maximum height of the profile peak
Rq	Quadratic mean deviation of the profile
Rv	Maximum valley of profile sink
LPM	Laser profilometer
STN	Slovak Technical Standard
EN	European Standard
MESH	Unit of abrasive grain size
h	Steepness of sample surface measurement
v	Shift speed of machine cutting head
3D	Three-dimensional image
P	Primary profile
W	Waviness profile
R	Roughness profile

Chapter 1
Current State of the Art

A number of adequate descriptions of 3D optical topographic methods are available at present [1–6]. 3D optical metrology represents a branch that has undergone significant development in the last decades especially owing to the development of computer technology, optical detectors, and light sources.

When selecting a particular optical topographic method it is inevitable to stem from several criteria which follow the assignment of the experiment. The criteria include requirements related to measurement speed, size of measurement field, method sensitivity, the precision of results, and financial intensity. Should the manufacturing of a sensor be part of an experiment, the selection of the method shall also include requirements regarding attendance qualification. As currently, the range of well-known and applied optical topographic methods is wide, it shall be possible to opt for the one which shall represent a compromise of these criteria.

1.1 Surface Roughness

Surface roughness is defined as a complex of surface irregularities with relatively short distances occurring as a result of applied component manufacturing technology. Roughness excludes surface defects formed only exceptionally (cracks, sinks, etc.) as well as defects caused by material failure, damage, etc.

The roughness of the machined area is generally defined as a slight irregularity of the machined surface caused by tool traces, uneven cutting, shaking, etc. In the case of non-machined areas, surface roughness occurs due to casting mold imprints, die blocks, and material deformation caused by the influence of compressive, tensile, flexural, and shear stress. Surface roughness is assessed according to the peaks and valleys of the area. It is detected visually, with a magnifier, with a microscope, through light interference, or by electric measurement.

J. Ružbarský, *Contactless System for Measurement and Evaluation of Machined Surfaces*, SpringerBriefs in Applied Sciences and Technology, https://doi.org/10.1007/978-3-031-08981-7_1

Roughness type and grade depend on machining method, physical and mechanical properties of machined material, cutting conditions, and especially on shift length and cutting speed. At all times the machined surface varies from the ideal geometrical shape. Zero roughness, i.e. ideal "smoothness" cannot be achieved in the case of technical surfaces.

Irregularities of machined components include a complex of deviations of the geometrical shape of the surface which can be classified from the technical and functional point of view as follows:

- Shape deviations refer to deviations of roundness, cylindricity, and flatness which are typical for the relatively low height of irregularity about a span the reason for which is incorrect gripping, deformation of a workpiece, inappropriate machine attendance, wear.
- Waviness is characterized by relatively precise shapes of irregularities similar to sinusoidal shapes occurring periodically with identical span and amplitude the cause of which is machine or tool shaking.
- Periodical roughness is formed mostly by irregularities whose shape and periodicity are evenly spread on the surface of the machined area. Their cause rests in tool shape, cutting conditions, sharpening, and planning.
- Random roughness is formed by irregularities whose shape and periodicity are randomly spread on the surface of the machined area. Their cause rests in random impacts during surface formation.

Shape deviations and waviness are referred to as surface macrogeometry, and roughness is referred to as surface microgeometry. Two terms are important when assessing the surface roughness:

- Surface imperfection,
- Surface structure.

Surface imperfections are represented by scissures, cracks, pores, corrosion, micro-cracks, etc. which occurred during manufacturing, storage, or surface function. These surface imperfections exclude surface structures from assessment.

The surface structure is periodical or random deviation from the geometrical surface which forms three-dimensional surface topography. When assessing surface roughness a middle profile line must be stemmed from, which represents a basic line and which [7]:

- Has a shape of the geometrical profile and divides the profile as follows: in the length range within the framework of which the roughness is determined the sum of squares of profile deviations is the lowest from the line that is referred to as a middle line of the smallest squares of the profile,
- Has a shape of the geometrical profile and divides the profile as follows: in the length range within the frame of which the roughness is determined the sum of areas bordered by the middle arithmetic line of the profile and the profile on both sides are identical.

- According to STN EN ISO 4287 standard the following standardized surface profile parameters can be distinguished: primary profile P, waviness profile W, and roughness profile R [8].

1.2 Machined Surface Characteristics

Irregularities occurring in manufacturing or machined areas are considered to be a surface. Quite often the area is viewed as a certain characteristic arrangement that is a result of mutual geometric and kinematic relations, tools, and workpieces accompanied by physical phenomena. Change of surface properties of the workpiece or change of surface character can be observed in dependence on conditions affecting machining, type of material, the material of a tool, etc. The tip of a tool is considered to be the most significant aspect influencing the final shape of the machined area. The shift represents other factors that characterize the relation between the tip of the tool and the workpiece [9].

Surface roughness is the property that influences service life and reliability of components, energy losses, wear resistance, tribological properties, etc. From the point of view of the technology of component machining the surface roughness nowadays represents one of the standardized criteria of their quality. Surface roughness considerably influences the development of physical and chemical phenomena in case of component functions, friction, the effectiveness of tightening, functional reliability, service life, and economy of machine and equipment operation [7].

1.2.1 Quality of Machined Surface Area

From the technological point of view of manufacturing, surface quality is referred to as the precision of dimensions, geometrical shape, position, and roughness of the surface. Surface roughness considerably influences the service life of equipment, reliability of machine run and accuracy of machine component operation, noisiness, running-in period, fatigue strength, wear resistance, corrosion resistance, etc. Due to the aforementioned, it is desirable to monitor the roughness of functional areas and evaluate the measured parameters [7].

The surface roughness of the machined components differs from the ideal roughness given by a drawing. The machined area is not ideally smooth. Micro-irregularities are present in the machined area. Force action of the tool a thin surface layer is deformed under the machined area. The type and grade of roughness depend on the machining method, and physical and mechanical properties, especially on shift length and cutting speed. Due to surface layer deformation and warming by the heat generated during machining, stress is formed, and physical and mechanical properties of the layer change [7].

Surface quality as an integrated feature of machine parts is characterized by the following [7]:

- Machined surface geometry,
- Deviations from the ideal condition,
- Physical and mechanical properties of surface layer,
- Hardness, strengthening and annealing stress,
- Physical and chemical condition of the surface.

The most significant and the most frequent quantity indicators extending the surface integrity are the following [7]:

- Machined area macrogeometry (shape deviations),
- Machined area microgeometry (roughness),
- Changes in physical and mechanical properties of surface layer,
- Annealing stress under the machined surface,
- Physical and chemical condition of the surface.

These indicators form preconditions for influencing fatigue strength, wear resistance, anticorrosion stability, the precision of fitting, etc. Moreover, they are of high significance for dynamically stressed parts and parts subjected to wear. From the point of view of surface quality, the micro-geometric deviations are of high significance, i.e. surface roughness [7].

1.2.2 *Theoretical and Actual Surface Roughness*

From the point of view of methodology, it is possible to distinguish between theoretical roughness and actual surface roughness [7].

Theoretical roughness can be geometrically determined if the following is met [7]:

- Machined material is considered to be non-deformable,
- The cutting edge of tool forms geometrical lines,
- System machine—tool-workpiece is fixed.

If the aforementioned preconditions are met, theoretical surface roughness can be determined or the maximum height of irregularities can be defined according to formulas corresponding to kinematic schemes of removed chips. Actual values of roughness characteristics differ from the calculated values. These differences are recorded by profile graphs where, for instance, the shape of the lathe-turning surface loses regularity contrary to the theoretical profile. The reasons for the change of the machined profile shape and for an increase of actual heights of irregularities when compared with the theoretical ones can be searched in material and technological factors [7].

The basis for surface scanning and evaluation is represented by the surface profile. A surface profile is an intersection between surface and plane. The plane can be situated in the direction of irregularities (longitudinal profile) and perpendicularly to the direction of irregularities (transversal profile). The surface profile is characterized by irregularities along the height direction and by irregularities along the length direction.

Surface geometry profile is a complex of irregularities with negligible spans formed during its machining. The machined surface contains traces left by the machining tool and the non-machined one has irregular imprints of the die casting mold, imprints of irregularities caused by the cylinder, etc.

The surface structure is assessed through reduction toward the cutting plane which is perpendicular to the surface. Thus the surface profile is achieved in the cutting plane (Fig. 1.1) which consists of a waviness profile, roughness profile, and shorter and longer elements of waves occurring on the surface. The individual elements are separated through filtration with the use of diverse types of filters [7].

According to STN EN ISO 4287 three types of surface profiles are distinguished (Fig. 1.2) [9]:

- Primary profile – P-profile,

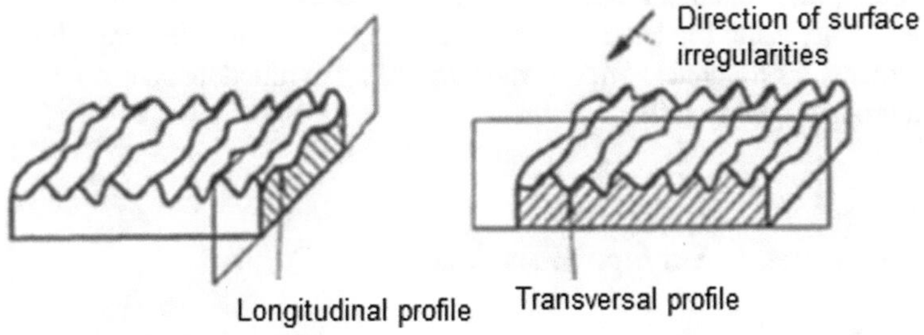

Fig. 1.1 Longitudinal and transversal profile of surface geometry [10]

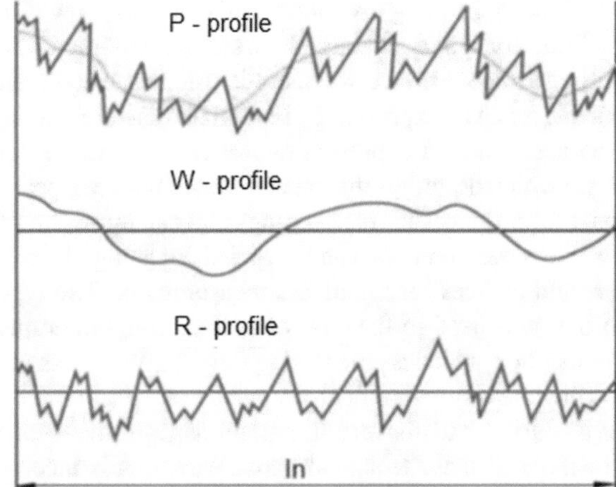

Fig. 1.2 Primary profile with its derived roughness and waviness profiles [9]

- Roughness profile – R-profile,
- Waviness profile –W-profile.

Suitable features and parameters, which are considered to be surface roughness criteria, are measures. Measurement of parameters and features is defined by STN EN ISO 4287-1,2; 4288; 5436 standards and a unit of measurement is μm.

Surface roughness is a technical quantity with a significant impact on friction in contact areas. It is assessed from the point of view of quantity and quality according to diverse criteria. Currently, contact profilometers and profile graphs are used for the quantitative assessment of the surface structure. However, new methods have appeared which bring several advantages. These digital devices precisely analyze surfaces and based on collected data they evaluate surface roughness parameters. Contrary to the past, it is rapid progress since several years ago only a visual check was applied in the case in which just a manufactured surface was compared with the calibrated sample. Also, optical measurement devices were used using a microscope the material surface was enlarged and consequently compared with an etalon [10].

Surface roughness is prescribed for all component areas in the drawing except for the areas for which roughness is not important, for instance, neck-downs, center holes of chamfers, and rounded edges. Roughness is not characterized by conventional threaded areas, drilled holes, counterbores and countersinks for screw heads, etc. Requirements for roughness are specified according to functional demands to assure desired product quality [10].

1.2.3 Geometric Surface Waviness

Geometric surface waviness is characterized by larger irregularities upon which roughness is layered. Frequently it is caused by the workpiece vibrations or deformations. Waviness is mostly ascribed to properties of a machine tool, for instance, unbalanced grinding disc, inaccuracy of lead screws, insufficient toughness, etc.

Contrary to surface roughness, surface waviness is measured as to the width of the entire surface. Surface waviness is usually periodic alike in the case of roughness, waviness can be expressed by several parameters. For instance, Wa stands for average waviness, and Wt expresses total waviness. Another parameter to be mentioned is Wsm which describes the mean distance between peaks of periodic waviness. There exist several measurement settings that can influence the results of these parameters, waviness measurement can be carried out using diverse devices including profilometers and devices for roundness measurement. The type of such devices is identical to the ones used in the case of surface roughness measurement, i.e. those are the contact or contactless measuring devices. Waviness measurement is not as frequent as roughness measurement yet waviness measurement (e.g. waviness measurement of a bearing ball) is very important as right the waviness increases vibrations and noisiness of these technical parts. Waviness is incorporated in ISO 4287 and ISO 16610-21 standards [9].

1.2.3.1 Surface Microgeometry

The machined area is formed as a trajectory wrap area of working movement of a machine cutting edge and from basic geometric areas given by a drawing, i.e. it significantly differs from the plane, cylindrical, or another surface. For instance, the longitudinally turned area is helical, the planed area consists of several side-by-side positioned grooves; the ground area is marked with scratches left by abrasive grains, etc. The machined area is characterized by microgeometry which has significant meaning from the point of view of the future function of the area [1].

Surface microgeometry is influenced by the following factors [1]:

- Plastic deformation at the moment of chip formation,
- Dynamic phenomena, i.e. vibration occurring in a technological system,
- Cutting conditions of the machining process,
- The friction of cutting wedge back against the machined area.

Actual areas have a certain deviation from nominal values. These deviations are the results of imprecise manufacturing.

Macrogeometry is influenced by the following [1]:

1. Imprecision of machining tool [1]:
    - geometric (is given by deviations from the prescribed reciprocal position of functional machine parts),
    - kinematic (is characterized by deviations of actual trajectory of machine mechanisms from the ideal one),
    - dynamic (is given by deviations of the reciprocal position of machine nodes when being loaded by cutting forces).
2. Imprecision of a tool (tool shape, cutting edge wear, cutting geometry, cutting resistance, imprecision caused by inaccurate dimension setting of the tool),
3. Imprecision of applied agents (imprecision of agents in determination of mutual position of tool and workpiece, workpiece deformation caused by clamping forces, deformation of inherent agent caused by cutting force action).

1.3 Quality Parameters

The suitable surface quality of components represents one of the preconditions of correct machine function and significantly influences the service life of components, for instance, the surface quality of switching contacts, the surface roughness of a potentiometer slide, etc. Surface quality refers not only to the geometric shape of the surface and the size of its irregularities but also relates to the physical and chemical state of the surface layer of the material. One of the main criteria in the case of surface quality and machinability assessment is represented by roughness.

1.3.1 Evaluation of Surface Character

Evaluation of surface character by standards stems from the profile method. The method evaluates the surface according to profile. The profile represents a line after cutting the actual surface defined by the area. In practice, such a plane is selected the normal line of which is perpendicular to the actual surface and disposes of the appropriate direction. Appropriate direction is the direction in a case in which higher values of surface character parameters (transverse profile) are detected unless the component drawing prescribes the direction as well. Surface roughness is measured according to several characteristics. The following characteristics are frequently measured [11]:

- The arithmetical mean deviation of profile R_a,
- Maximum height of the profile roughness R_z,
- The quadratic mean deviation of profile R_q [11].

The arithmetical mean deviation of profile R_a (Fig. 1.3) occurs frequently in drawings. It is the arithmetical mean value of absolute deviations of profile within the range of primary length L [11].

$$R_a = \frac{1}{\mathrm{lr}} \int_0^{\mathrm{lr}} |Z(x)|\, \mathrm{d}x \tag{1.1}$$

with

lr measured section length,
x profile line segment readout on a middle line,
$y(x)$ function describing profile [11].

Maximum height of profile roughness R_z: (Fig. 1.4) height characteristics of surface roughness is determined by the distance between the line of profile peaks and the line of profile valleys within the range of primary length [11].

$$R_z = R_p + R_v \tag{1.2}$$

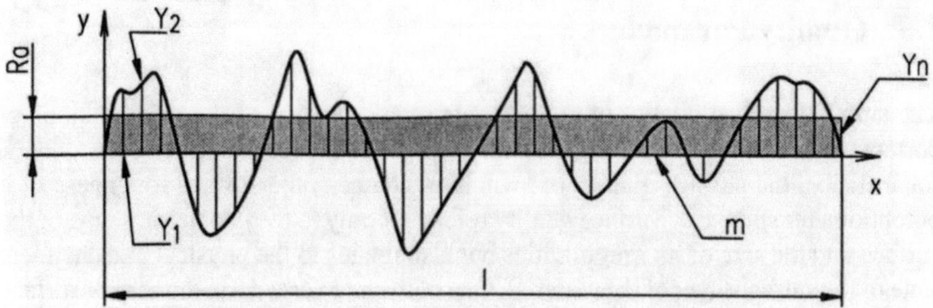

Fig. 1.3 Arithmetical mean deviation of profile R_a [11]

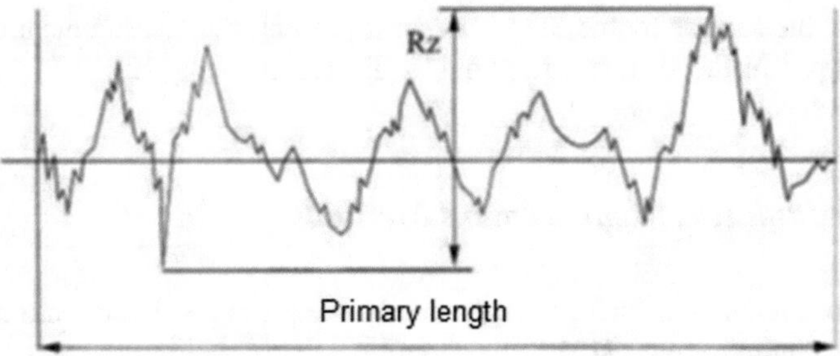

Fig. 1.4 Maximum peak of profile roughness R_z [11]

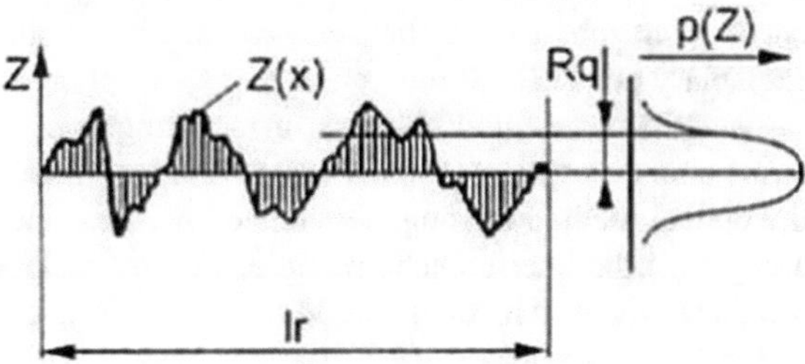

Fig. 1.5 Quadratic mean deviation of profile R_q [12]

with

*R*p the height of the maximum peak within the range of primary length
lr, R_v depth of the largest valley within the range of primary length lr [11]

The quadratic mean deviation of profile R_q (Fig. 1.5) shows higher values than the R_a value; it is more sensitive to undesired surface peaks and valleys. This parameter is used especially in the optical industry. [12].

$$R_q = \sqrt{\frac{1}{\mathrm{lr}}} \int_0^{\mathrm{lr}} Z^2(x)\,\mathrm{dx} \tag{1.3}$$

with lr—measured section length [12].

1.4 Measurement of Machined Surface Roughness

Check of surface quality currently represents a significant part of surface production in case of any technology used for its formation. Since 1930, when the surface topography measuring devices were developed, remarkable progress has been made in the field of measuring methods and equipment. A significant advance was the application of digital methods in the 1960s which also represented the possibility to evaluate the surface by the 3D method. At present, the measurement of surface roughness parameters can be carried out by diverse methods [12].

1.4.1 Roughness Measurement Methods

The methods applied in surface roughness checking can be divided into qualitative ones and quantitative ones [12].

The qualitative methods are based on a subjective comparison of the checked surface with the exemplary one the roughness of which is given. Only surfaces machined with the same or at least with a similar method can be compared. The check result proves the checked area to be smoother or rougher contrary to the exemplary one [12].

Quantitative methods express surface roughness numerically. These devices are based either on optical (contactless) or contact measurement methods. In the case of the optical method of roughness detection, the principle of light cutting is applied along with the interference principle. Current methods used for machined surface roughness measurement are based on diverse physical principles [12].

1.4.2 Roughness Measurement Devices and Methods

Individual measurement devices can be divided into three basic categories as follows [12]:

- Contact measurement devices,
- Contactless measurement devices,
- Other measurement devices.

1.5 Contact Methods of Roughness Measurement

Contact measurement (Fig. 1.6) of surface roughness represents the most widespread method of roughness measurement, especially in mechanical engineering practice.

The advantage of the method rests on indirect measurement and the possibility of use for any surface type. In the case of this roughness detection method, a measurement touch probe with a small diameter of a stylus (2 μm up to 10 μm) moves along the surface which is part of the sensor based on the induction or piezoelectric principle. The measurement touch probe performs motoric movement along the measured surface [13]. Table 1.1 shows the advantages and disadvantages of contact measurement methods.

A contact measurement device is defined as a device examining respective surfaces with a stylus and calculating surface profile parameters. Measuring with such a

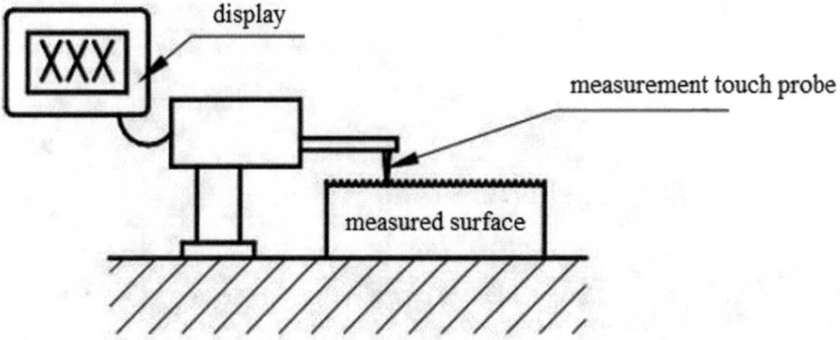

Fig. 1.6 Scheme of contact measurement device [13]

Table 1.1 Advantages and disadvantages of contact method of surface roughness measurement [14]

Advantages	Disadvantages
Direct measurement of surface height irregularities compared with the reference plane	Surface destruction in case of soft materials
High definition in the direction perpendicular to the middle plane of the surface	Relatively slow stylus movement along the measured surface
Broad measuring range in the direction perpendicular to the middle plane of the surface	It does not allow continual check and regulation of surface quality
Possibility of measuring in an impure environment	Measurement cannot be remotely controlled
Definition of standardized parameters of surface topography based on precondition of use of this method for their measurement	Only 2D measurement (3D measurement is time-consuming)

device is fast and simple. There exist a large number of contact devices and due to development, it is inevitable to measure more inaccessible surfaces therefore the use of portable contact measurement devices increases [15].

1.5.1 Measruement System MAHR MarSurf PS1

This device (Fig. 1.7) represents unlimited mobility. Device operation is simple and does not require any training. The device can be used both in horizontal and vertical positions. Owing to the large and transparent display the work with the device is simple [15].

Device MarSurf PS1 (Table 1.2) for evaluation of surface character is as standard equipped by a movable unit, a standardized sensor, an inbuilt accumulator, integrated roughness standard, a height adjustment system, a protection cover of a sensor, a

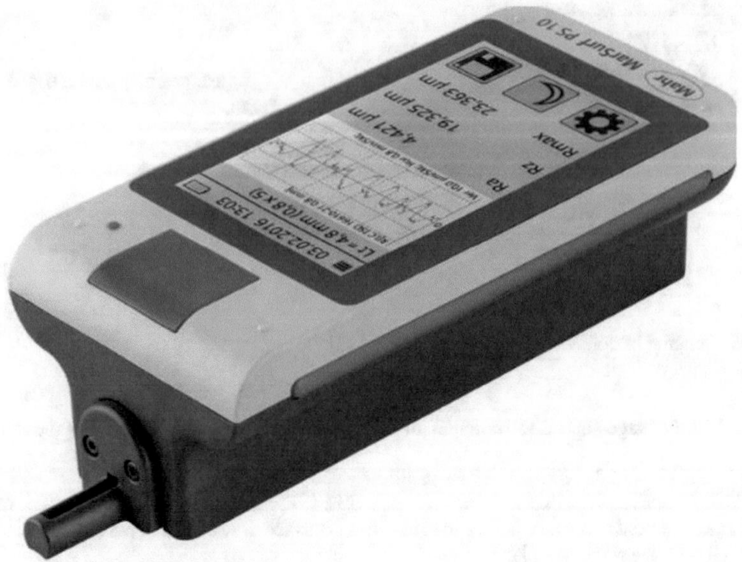

Fig. 1.7 Device MAHR MarSurf PS [15]

Table 1.2 Technical parameters roughness tester MarSurf PS1 [14, 15]

Unit of measure	Metric/inch [mm, μm/inch, μinch]
Measuring principle	Contact method
Number of roughness parameters	25 (R_a, R_q, R_z, and other)
Measuring range	350, 180, 90 μm
Definition—range	0.01 μm
Measuring speed	1 mm/s

charger/a network adapter, an instruction manual, a strap shoulder bag and a mini USB cable [15].

1.5.2 Measuring Instrument CarlZeiss Handysurf E-35A

A small portable roughness tester (Fig. 1.8) is characterized by precision and simple operation. The tester disposes of inherent data memory and instantly evaluates and displays data. The device can be applied especially in case of input check, during manufacturing, and manufacturing control. Handysurf E-35A is a device for measuring in hardly accessible areas. Diverse applications utilize various sensors which can be easily and quickly replaced [16].

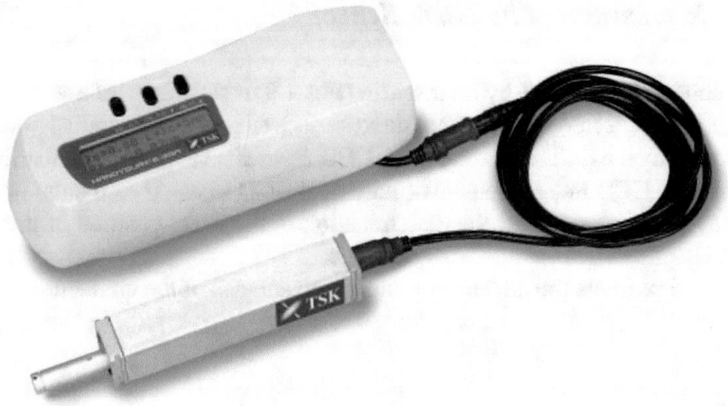

Fig. 1.8 Roughness Tester Handysurf E-35A [16]

Table 1.3 Technical parameters of device Handysurf E-35A [14, 16]

Unit of measure	Metric/inch [mm, μm/inch, μinch]
Measuring principle	Contact method
Number of roughness parameters	18 (R_a, R_z, and other)
Measuring range	40, 160 μm
Definition—range	0.02 μm
Measuring speed	0.6 mm/s

The device (Table 1.3) is supported by ISO, DIN, CNOMO, JIS, and ASME standards. At the same time, it supports seven languages. Measurement can be carried out horizontally and vertically and from below as well. For more flexible use the movable unit can be detached from the displaying unit and an additional fixing agent can be applied instead. The measured values, parameters, and points of the profile can be transferred directly to a PC through a serial interface (RS 232C) [14, 16].

1.6 Contactless Methods of Roughness Measurement

Contactless (optical) methods of measurement replace contact measurement of surfaces sensitive to mechanical damage, soft materials, etc. Contactless measurement excludes damaging of the measured surface. The checked surface is monitored by a focused measuring head in the program-controlled setting which is fast and simple. The disadvantage of optical methods rests in the fact that such methods are indirect [17].

1.6.1 Measurement by Laser Sensor

The principle of measuring by laser sensor (Fig. 1.9) rests in a ray of a semiconductor laser which is reflected from the measured surface to a receiving optical system. The ray is focused on a CCD scanning field. CCD assures a value of light distribution in the ray point. CCD image elements (individual CCD scanned elements) in the area of the ray point are used for the determination of the exact position of the focused point [18].

Table 1.4 presents the advantages and disadvantages of the contactless measurement method.

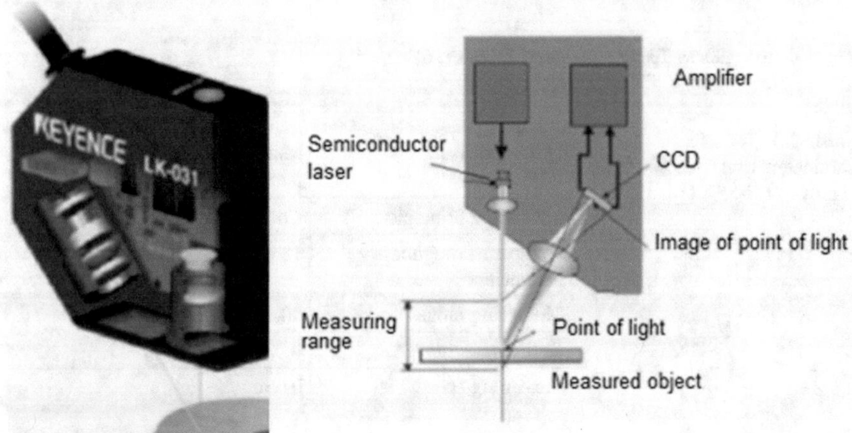

Fig. 1.9 Scheme of laser sensor [18]

Table 1.4 Advantages and disadvantages of contactless measurement of surface roughness [13, 14]

Advantages	Disadvantages
Possibility of continual check and control of surface quality	Measurement of surface roughness parameters is indirect
Measurement repeatability	More complicated means of result interpretation
Contactless character and non-destructiveness	
Possibility of measurement from a longer distance	

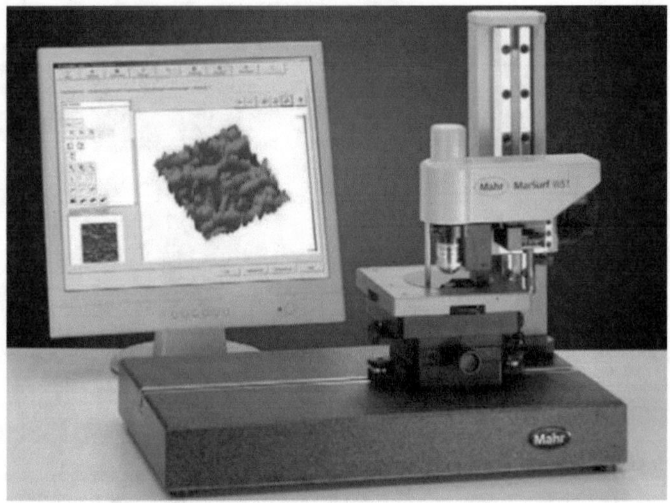

Fig. 1.10 Device MarSurf WS 1[19]

1.6.2 Measurement System MAHR MarSurf WS 1

Optical sensor WS 1 (Fig. 1.10) serves for contactless measurement of surface structure. The device works on the principle of white light interferometry. The technology allows fast and highly precise and contactless capturing of surface topography of various materials [19].

A high vertical definition of 0.1 nm allows for recording even the finest structures. The device (Fig. 1.11) is applicable in a measurement center as well as in a manufacturing environment [19].

MarSurf WS 1 and its innovative measurement evaluations allow an analysis of reflective and rough workpieces. Due to its high vertical definition (Table 1.5) it allows for measuring the surface roughness of optical elements such as lenses and mirrors [20].

1.6.3 Measuring Instrument Taylor Hobson CCI HD

Device Taylor Hobson CCI HD (Fig. 1.12) is a contactless optical profilometer for the measurement of surface profile and layer thickness. A robotic vertical scanner without piezoelectric elements saves high service costs. Simple operation limits the number of operator errors. Speed and exceptional sensitivity make CCI HD an ideal instrument for research and development as well as for quality assurance in the manufacturing process [21].

Fig. 1.11 Connection diagram of MarSUrf WS 1[19]

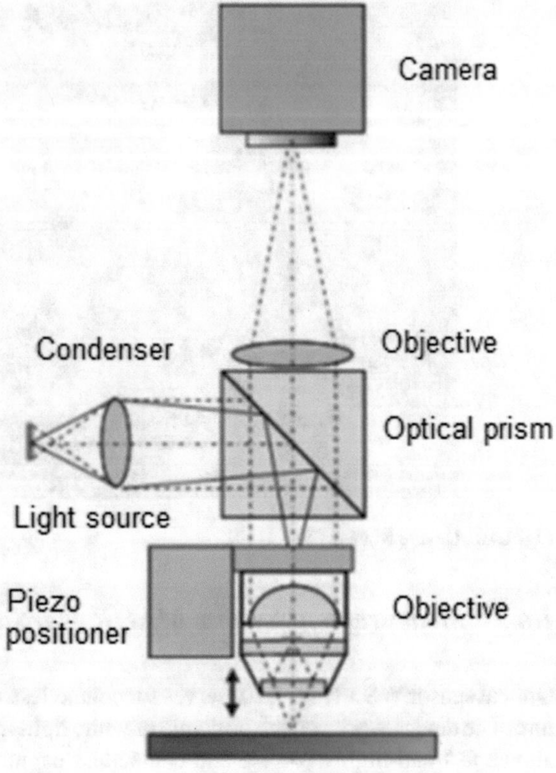

Table 1.5 Technical parameters MarSurf WS 1 [14, 20]

Unit of measure	Metric/inch [mm, μm, nm/inch, μinch, ninch]
Measuring principle	3D optical interferometry
Objectives	10x, 20x
Measuring speed	1.5–15 μm/s
Measuring range	100–400 μm
Definition—range	0.1 nm

CCI HD is ready to keep pace with the high professional level of scientists and research staff and to cover the majority of challenging requirements (Table 1.6) related to measurements in fast-developing branches [14, 21].

Fig. 1.12 Device TaylorHobson CCI HD[21]

Table 1.6 Technical parameters of device TaylorHobson CCI HD [14, 21]

Unit of measure	Metric/ inch [mm, μm, nm /inch, μinch, ninch]
Measuring principle	3D contactless measurement, coherent correlation interferometry
Objectives	2.5x, 5x, 10x, 20x, 50x, 100x
Measuring speed	36 mm/s
Measuring range	2 mm
Definition—range	0.01 nm

1.7 Laser Profilometry LPM

The basic method for the evaluation of the geometric shape of objects is a contactless measurement of 3D profile—laser profilometry—LPM. Optical methods allow the measurement of geometrical parameters of objects which are complicated or impossible to be measured by standard measurement methods. This method represents a much faster, more precise, and more reliable solution contrary unlike manual measurement systems [22].

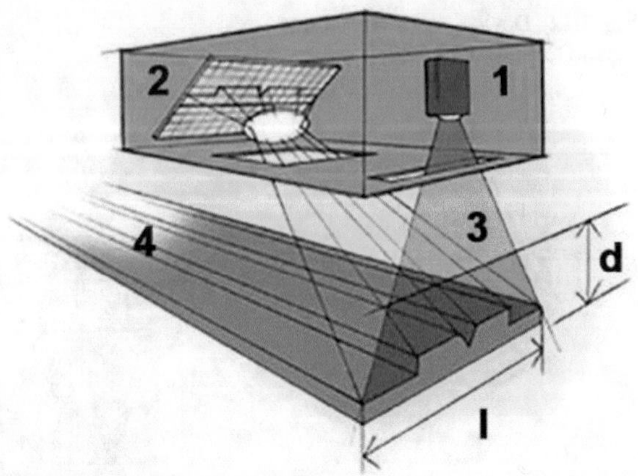

Fig. 1.13 Principle of laser profilometer. 1—CCD camera, 2—laser light source, 3—range of laser light scanning by a camera, 4—measured surface, d—working range of profilometer, 1—primary measurement length [22]

The compact profilometer (shown in Fig. 1.13) serves for contactless optical measurement of the 3D profile along with the defined cutting profile. The triangulation principle as a well-known method is used for this purpose. In the case of this method, the laser ray is projected onto the measured object. The laser ray trace is scanned at an angle by a digital camera. Each scanned image serves for the evaluation of the actual 3D profile [22].

The following ranks among the biggest advantages of contactless laser profilometry LPM [23]:

- Wear of measuring instrument and damage of measured object surface is avoided during measurement,
- Measuring of absolute roughness of the measured surface, contrary to contact methods in case of which measuring stylus cannot precisely enter into cracks or micro-irregularities,
- Repeatability of measurement,
- Working range of measurement allows measuring even more curved surfaces contrary to the contact method,
- Evaluation of several parameters of roughness and waviness,
- Measurement of the entire part of the sample surface, not only of a single straight line.

The following ranks among the biggest disadvantages of contactless laser profilometry LPM [23]:

- Problem with the measurement of polished surfaces,
- Problem to measure multi-layered surfaces,
- The higher purchase price,
- Trained attendance [23].

1.7.1 Laser

A laser is a source of monochromatic coherent light formed when a light amplifier is placed into an optical resonator tuned up to a particular wavelength. Laser light is the brightest existing light, even brighter than the sun. In a laser a thin beam of rays is formed which is intensive enough to burn a hole in steel. This beam of the ray is direct and thin and can be precisely directed toward the mirror positioned on the Moon, which equals a distance of 384,401 km. At the same time, it is direct enough for diverse precise measurements, i.e. more suitable for precise measurement than any other ruler. Moreover, such light is coherent, which means that it contains only identical wavelengths (identical color) and the waves are in perfect harmony with time [24] (Fig. 1.14).

Laser light: Inside the laser, there is a tube containing a gas mixture such as helium and neon, liquid or solid crystal, for instance, ruby. Gas lasers (e.g. argon laser) produce low-intensity ray beams intended for fine operations, for instance in eye surgery. Liquid fluorine hydride in chemical lasers is used for the production of intensive rays intended for weapons [24].

Two electrodes produce an electric spark end in the tube. The spark supplies atoms of emitting material with sufficient energy to get excited and to radiate photons—a subtle flash of light. Photons radiated out of the laser material are emitted in all directions and collide with other atoms which radiate photons as well. In a short

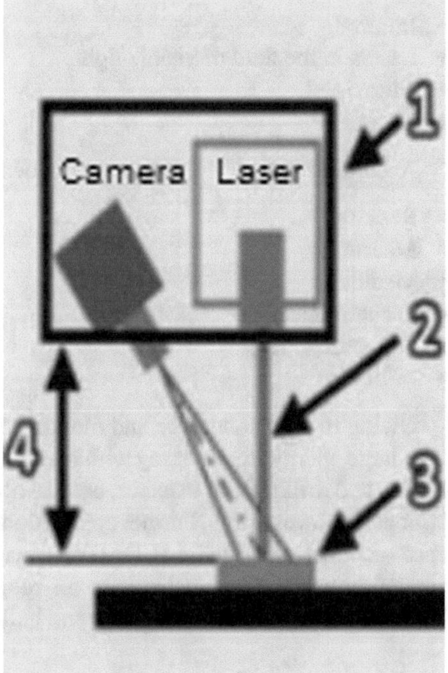

Fig. 1.14 Roughness measurements with laser

period, billions of identical photons are produced moving fast from one end of the tube to the other [24].

At each end, a mirror reflects photons flittering back and forth within the tube so repeatedly they flitter from one end to the other and emit further photons. One of the mirrors' constructions allows transmission of a certain amount of photons so that they get accumulated behind it. Once a sufficient number of photons is accumulated the laser emits an intensive ray beam. Some lasers emit continuous ray beams. High-performing pulsed lasers emit ray beams at regular intervals [24].

According to the type of energy, pumping lasers can be categorized as follows [24]:

- Optically pumped lasers (with discharge lamp, with another laser, with sunlight and radioactive emission),
- Electrically pumped lasers (with collisions in an electric discharge, with a bunch of charged particles, with electron-injecting, with the interaction of the electromagnetic field with a bunch of charged particles),
- Chemically pumped lasers (with chemical bond energy, with photochemical dissociation, with energy exchange between molecules and atoms),
- Thermodynamically pumped lasers (with air/gas heating and cooling),
- Nuclear pumped (with reactors, with a nuclear explosion,
- From the point of view of working mode, the lasers can operate continuously or impulsively.

Lasers can be divided according to emitted wavelength as follows [24]:

- Infrared,
- Lasers in the field of visible light,
- Ultraviolet,
- X-ray.

According to their use lasers are divided as follows [24]:

- Research,
- Measuring,
- Medical,
- Technological,
- Energetic,
- Military.

Owing to high coherence and monochromatic character it is possible to concentrate large energy in a laser ray within a small area which is a base of use for material cutting and drilling. For instance, conical holes for engine valves can be produced by laser at the lowest costs. Further applications use insignificant divergence and coherence—optical data media (CD, DVD, magneto-optical discs), measuring devices applied in science and technology for roughness measurement. At the same time, they are employed in the medicine and building industry [24].

1.7.2 Laser Light Color and Performance

Red lasers mostly operate on the wavelength of 650 nm. They rank among the cheaper ones and their production is less demanding. Red lasers appeared in the market among the first produced lasers. When illuminating, the finite point can be seen yet their ray is not seen only in the case of the stronger ones. They are commonly used as presentation pointers incorporated even in pens or wireless power-point remote controls [24]. Table 1.7 shows the performance specifications of these lasers.

Green lasers mostly operate on the wavelength of 532 nm. Green lasers are slightly more expensive. The wavelength of this color can be caught by the eye best, i.e. these green lasers have the brightest ray contrary, unlike other colors. They are frequently used as pointers at night or as an exquisite light effect at discos [24]. Table 1.8 shows the performance specifications of these lasers.

Blue lasers operate on the wavelength of 445 nm. The price of these lasers is currently the highest as their production is the most costly and they represent a minor line of products. Yet the attractiveness of these lasers is due to the aforementioned rather high. Blue laser is a bit of luxury for the users. Illuminating properties are similar to those of green lasers. But the human eye can catch green rays better than blue ones. Blue lasers dispose of similar applications as the green ones (this type of

Table 1.7 Red laser [24]

Effects of 650 nm red lasers				
Wavelength (mW)	5	50	100	200+
Visible finite element	●	●	●	●
Burns thin black plastics		●	●	●
Blows out colored balloons		●	●	●
Lights a match		●	●	●
Cuts a black antistatic tape			●	●
Laser ray visible at night				●

Table 1.8 Green laser [24]

Effects of 532 nm green lasers					
Wavelength (mW)	5	30	50	100	200+
Visible finite element	●	●	●	●	●
Laser ray visible at night		●	●	●	●
Burns thin black plastics		●	●	●	●
Blows out colored balloons			●	●	●
Lights a match			●	●	●
Cuts black antistatic tape				●	●
Laser ray visible during a day					●

Table 1.9 Blue laser [24]

Effects of 445 nm blue lasers					
Wavelength (mW)	5	100	200	1000	1200+
Visible finite element	●	●	●	●	●
Blows out colored balloons	●	●	●	●	●
Lights a match	●	●	●	●	●
Cuts black antistatic tape		●	●	●	●
Burns rubber			●	●	●
Laser ray visible during a day				●	●
Lights a match (the wooden part without sulfur)					●

laser is used in our LPM) [24]. Table 1.9 presents the performance specifications of these lasers.

Purple/UV lasers frequently operate on the wavelength of 405 nm. As to price these lasers are between green and blue. Many times these lasers rank among the blue ones yet their wavelength is slightly shorter and their visibility borders the spectrum of our eyesight. Purple lasers are UV lasers. Properties of UV radiation are well-known for instance due to the determination of authenticity of banknotes in banks. That is caused by the fact that in case of reflection of UV rays from a white object it starts to illuminate [24]. Table 1.10 presents the specifications of these lasers.

Infrared/IR lasers mostly operate on the wavelength of 808–980–1024 nm. These lasers are the cheapest ones of all. Their wavelength cannot be caught by the eye or only slightly because it is a bit longer than the highest point of the eyesight spectrum. The border between the visible spectrum and IR radiation has not been precisely defined since it stems from human eye sensitivity. The longer the wavelength is, the lower the probability of finite point visibility is. Their application is for instance in optical fibers which guarantee the high speed of information transfer. Their infrared radiation is also used in several wireless remote controls such as IR LED which unlike the red laser serves for communication with the receiver, not for points. However,

Table 1.10 Purple laser [24]

Effects of 405 nm purple lasers				
Wavelength (mW)	50	100	200	500+
Visible finite element	●	●	●	●
Blows out colored balloons	●	●	●	●
Lights a match	●	●	●	●
Cuts black antistatic tape		●	●	●
Burns rubber			●	●
Laser ray visible at night				●

Table 1.11 Infrared laser [24]

Effect of 808 nm infrared laser				
Wavelength (mW)	200	500	1000	2000+
Burns thin black plastics	●	●	●	●
Blows out colored balloons	●	●	●	●
Lights a match	●	●	●	●
Cuts black antistatic tape	●	●	●	●
Visible finite point		Very poor visibility	Poor visibility	●

they can be applied to many laser fans due to their high performance for a relatively lower price [24]. Table 1.11 presents the performance specifications of these lasers.

1.8 Basic Components of Experimental Set-Ups of Optical 3D Methods

The scheme of the majority of experimental set-ups of optical 3D methods is very similar. A measured object, an optical detector, and an optical source are always present in a set-up (for easier adjustment and calibration of experimental set-up a light source is often used, i.e. optical source illuminating on the wavelength of 380–760 nm). Light sources are projected by a so-called optical probe which is a case of optical structure impinging on the measured object surface. The individual methods differ from each other in the type of this optical structure and the method of its interaction with the measured surface [24].

The rapid development of microelectronics and material physics in the last decades led to the broadening of the offer of light sources applicable in the case of experimental set-ups of optical 3D measuring instruments [26]. The laser used before is nowadays replaced by laser diodes which dispose of sufficient coherence length for the use of incoherent topographic methods and they have relatively low purchase prices. Spatial modulators and especially modern data projectors represent the basis for current incoherent 3D methods. Their main advantage stems from variability, i.e. the projected optical structure can be any image created by a computer. Further chapters describe basic types of optical structures used in the case of optical 3D measurement methods [25].

1.8.1 Point and Linear Optical Structures

The simplest optical structure to be applied is a point. In the majority of cases, the structure is realized through laser, i.e. by a laser diode with suitable optical elements. In such an event the project point disposes of sufficient light intensity, small dimensions, and Gaussian intensity curve which is convenient during the detection of coordinates of its center in case of the evaluation process. The disadvantage of this optical structure rests in insufficient information extraction when a result of a single measurement cycle is the topographic deviation of only one point of the measured object surface at coordinate (x, y) which corresponds with the geometrical center of the surface area of the lit optical structure. To gain complete topographic information on the object, i.e. on all coordinates (x, y, z) the source of light or the measured object must be moved using which all the coordinates (x, z) shall be covered. Another alternative is to select such density of covered points which would allow determining at least approximately the unmeasured part of the measured object surface with precision corresponding with experiment assignment [25].

Another relatively simple optical structure is a linear optical structure formed by light trace one coordinate of which remains constant along the entire length whereas the other changes coherently so that it covers the whole measured object in a single dimension. Linear optical structure is mostly carried out with laser or laser diode with a front-end cylindrical lens. Contrary to point projection the information extraction is much higher because the result of a single measurement cycle is dilations corresponding with all points of the axis in case a light probe changes continuously. Suitable change of axis coordinate in case of which linear coordinate of the light trace is constant and approximation of measured surface rest contributed to the achievement of complete topographic coordinates (x, y, z). These two optical structures are mostly used in the method referred to as laser triangulation (Chap. 1.9.1) [25].

1.8.2 Planar Optical Structures

Planar optical structure refers to a structure that is in a plane intersection perpendicular to the projecting axis of the projector and at the same time is two-dimensional. It means that it is projected in two axes. The planar optical structure can be realized by several methods. One of the methods includes interference in the case of which laser beam realized by laser or laser diode reflected from a plane-parallel plate. The optical structure formed this way consists of a system of parallel interference planes. Such light structure is used for instance in moiré methods (Chap. 1.9.5) in cases in which intensive sinusoidal development is applied. A further example includes optical structures formed by the spatial modulator. The light produced by the light source passes through the modulator which can be programmed when being connected to a computer and which generates desired light structure by phase or amplitude adjustment of the passing light beam. The advantage of such a trace is considerable variability. On the other hand, the disadvantage rests in modulator purchase costs, the necessity of an external light source, and the relatively low pixel definition of modulators [25].

The first-rate source of planar optical structures is represented by modern data projectors, especially the ones based on DLP technology (Digital Light Processing). These contain a light source the light ray of which impinges onto rotating colored disc or is distributed along with basic color with an optical prism. Such beam consequently impinges on a DLP chip comprising of a hundred thousand or even millions of small mirrors fixed to a joint. These mirrors are electronically controlled and according to the color and intensity of desired pixel corresponding with the particular mirror, they reflect light to or off the output objective. These mirrors make even 1024 moves per second. The data projectors have high definition, high contrast of the projected image, sufficient intensity, and can project any image created by the computer. They are used for instance in Fourier profilometry (Chap. 5.2), phase shifting profilometry (Chap. 1.9.3), or in the case of moiré methods (Chap. 1.9.5) [25].

Contrary to point or linear optical structures the planar ones have a great advantage. During one measuring cycle the entire measured object surface is measured. Although in general it can be stated that the precision of measurement of optical topographic methods using these planar light structures is lower than in the case of methods using point or linear structures, the methods offer complete coordinates (x, y, z) without the need of approximation and time demand factor is many times lower. Some of these methods can measure even in actual time.

The most frequently used planar optical structures are linear grids, sinusoidal grids, checkboard, etc. [25].

1.8.3 Intensity Optical Structures

This optical structure is very simple. In most applications, it comprises the spherical or plane light wavefront. Two basic principles of its use exist. It is used in the case of object illumination to create shades due to a curved measured surface (method referred to as shape from shading) when the measured object profile is calculated based on intensity and size of shade or according to their laser beam modified by suitable optical elements. This beam illuminates both measured and reference objects which can be formed by a flat or spherical mirror, lens, or by the interface of glass—air. These two intensity optical structures mutually interfere and thus interference image is created which is further evaluated and converted to topographic deviations. This principle is used by coherent topographic methods such as interferometry with white light (Chap. 1.9.6) or Fizeau interferometry (Chap. 1.9.7) [25].

1.8.4 Applied Optical Detectors

Optical detectors are designed for the detection of the position or deformation of the projected optical structure. This position change or deformation is caused by the reciprocal location of the optical source, the measured object surface, and the optical detector. Calculation based on calibration or according to the used method principle determines a particular deviation from such change or deformation. Linear optical detectors are used to a lower degree. Such detectors comprise a single series of optical sensors and they are used for instance in laser rangefinders, generally in such applications in the case that it is inevitable to determine position change of point or linear optical structure in one direction [25].

The majority of optical topographic methods use a camera as an optical detector. The basis of modern cameras used in current topographic methods is CMOS or CCD chips [27]. These cameras allow to record an image and save it in a computer memory where the image can be analyzed digitally and according to the principle of a particular method, the analysis can be employed in the evaluation of topographic deviations [25].

1.8.5 CCD Detectors

CCD chip (Charge-coupled device—a device with bound charges) was invented at the end of the 1960s. At the beginning of the 1990s, the technology made rapid progress, and nowadays CCD sensor is commonly used in digital cameras, cameras, barcode readers, scanners, etc. A flat CCD chip comprises the negative electrode, semiconductor, thin layer of silicon dioxide, and a network of positive electrodes. The layer of silicon dioxide is an insulant separating electrodes from semiconductors and preventing electrons from approaching the electrodes. The free movement of electrons along the chip is hindered by vertical potential barriers. Horizontal electrodes comprise a system of so-called "potential wells" out of which each represents one pixel, i.e. least observable image area. The principle of CCD chip activity is based on a photo effect which means the ability of impinging photon to release an electron out of the semiconductor. If a higher number of photons impinges on the semiconductor, then more electrons in the semiconductor get excited. Released electrons are attracted by positive electrodes and so-called "positive holes" are attracted by negative electrodes (Fig. 1.15) [25].

During image scanning a mechanical camera shutter is open and impinging light excites the electrons in the semiconductor. The number of electrons bound to one positive electrode is given by several photons impinging on each element of the chip

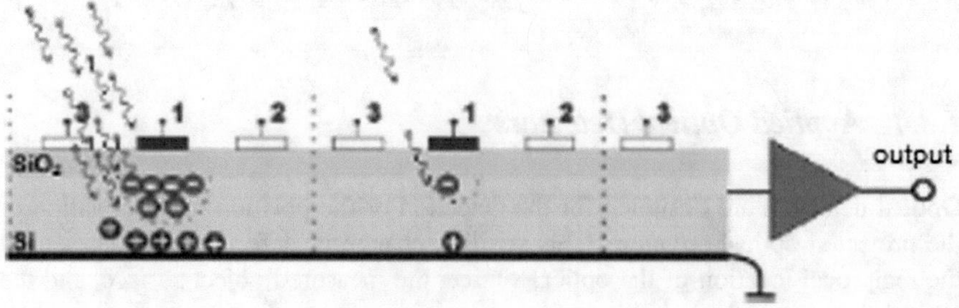

Fig. 1.15 Potential well of CCD chip [24]

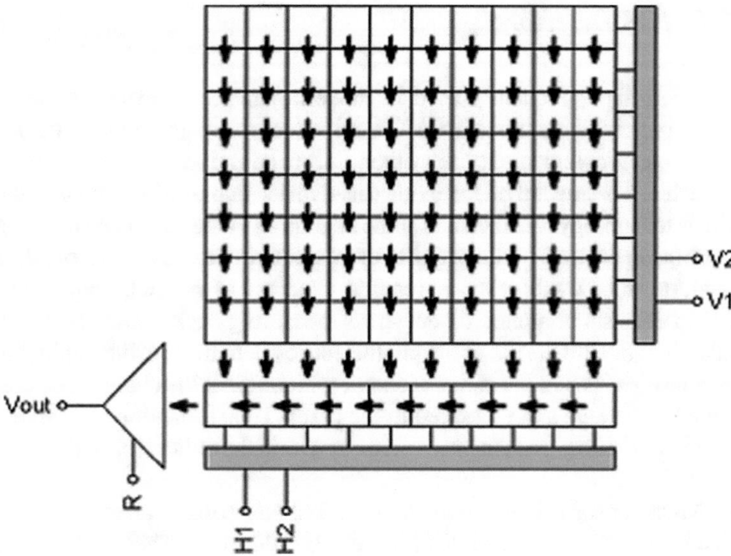

Fig. 1.16 Scheme of CCD sensor [28]

matrix. When the shutter has closed the change of the voltage of horizontal electrodes causes a shift of this bunch of electrons (i.e. charges) along with the chip matrix columns toward the linear CCD chip at their end where these charges are shifted by another set of electrodes toward the image amplifier in case of which a charge is transformed to voltage (Fig. 1.16). The magnitude of this voltage is measured and transformed to the intensity of light caught by each pixel because a direct proportion exists between these two quantities [25].

The cameras with black and white CCD chips are convenient for the majority of optical topographic methods. However, apart from intensity, it is inevitable to gain information on the color of the image pixels; there even exist modifications of CCD chips for scanning of color images. These chips´ pixels dispose of filters. One pixel of the final image is then created by an area of several pixels with diverse color filters the color of which is determined by the ratio of charges having impinged onto the individual pixels when they once passed through the color filter. Basic parameters of CCD chip quality are its definition (overall number) of pixels), sensitivity, dynamic range, and noise. Increasing requirements related to these parameters increase the purchase price of a camera equipped with the CCD chip. A suitable compromise between the parameters and purchase costs must be specified according to the experiment assignment [25].

1.8.6 CMOS Detectors

The other chip type frequently used in modern cameras in experimental set-ups of optical topographic methods is CMOS chips (complementary metal-oxide semiconductors—they complement each other, i.e. metal-oxide-semiconductors), i.e. a control electrode is situated on the oxide (insulant) which is located on the semiconductor. This technology was invented at the beginning of the 1960s and has been used since that time in the case of the majority of integrated circuits. Apart from the aforementioned, it can be used for image records. Contrary to the CCD chip, a pixel of a CMOS chip consists of a photodiode, stress measuring device, address, and output electronic circuit. With the aid of the photoelectric phenomenon, the impinging light again excites electrons in the semiconductor which forms the charge. The charge is transformed to voltage and measured in the pixel. Transformation of charges of all pixels to corresponding intensity is carried out within the entire chip at the same time [25].

The CMOS sensor is equipped with a vertical and horizontal address collecting bar which means that any individual pixel can be addressed. Therefore the intensity of all pixels does not need to pass through neighboring cells to the linear register (as happens in the case of CCD chips) through which the charge passes toward the analog and digital transducer. Owing to this architecture CMOS chips are faster than CCD chips and the energy consumption is incomparably lower. Contrary to CCD chips they are cheaper because any production line for integrated circuits can be used for their manufacturing. The disadvantage rests in the fact that due to the presence of electronic circuits in each pixel their overall area scanning light intensity is smaller than in the case of a CCD chip disposing of identical dimensions. At the same time, the noise intensity is higher in the case of CMOS chips yet their current rapid development gradually eliminates this disadvantage. Camera selection with CCD or CMOS chips depends on qualitative features given by experiment assignment and by suitability for a particular experimental set-up [25].

1.9 Overview of Applied 3D Methods

This chapter focuses on the description of the basic types of optical topographic methods—both coherent and incoherent methods. The chapter contains methods using all basic types of optical structures and principles which represent the basis of the majority of other topographic measurement methods. The selection of methods has been carried out in such a fashion which would allow comprehension of the complex issue of modern optical topography. It represents the methods widely used at present as well as methods that formed the historical basis for the majority of currently applied optical topographic methods [25].

1.9.1 Laser Triangulation

Laser triangulation, many times referred to as 3D scanning profilometry, is an incoherent optical topographic method using a point or linear trace as an optical probe. Owing to its advantages the trace is in the majority of cases realized with laser or laser diode and apart from use in rangefinders it is linear almost at all times [13, 29–31].

The principle rests in the projection of linear laser trace to the measured object and detection of its deformed image from the measured surface to the computer by the camera. The light source, the measured object, and the camera form a triangle the parameters of which (camera and light source distance from the reference plane, their angle about reference plane) is known. The scheme of experimental set-up is shown in Fig. 1.17 [13, 29–31].

The searched topographic deviation Δr is the distance of each point of the measured surface, the coordinates of which correspond with pixels of the camera chip matrix, from the reference plane. This topographic distance is described by the following relation [31]:

$$\Delta r = \frac{\Delta u}{b + a\Delta u} \tag{1.4}$$

with

Δr searched topographic deviation,

Δu number of pixels per camera chip matrix by which the point image shall be shifted during projection to the reference plane and the measured point,

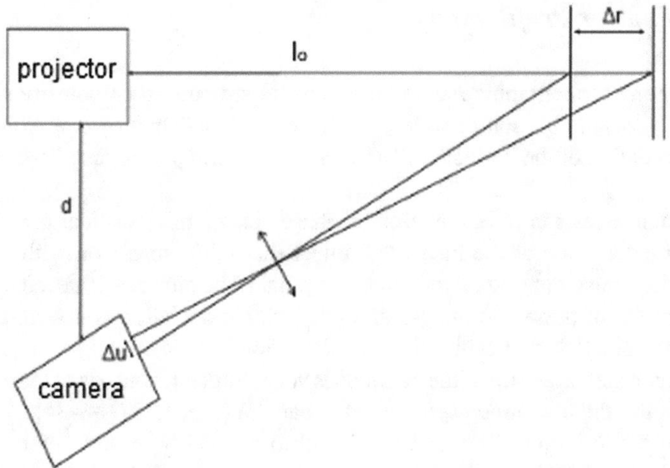

Fig. 1.17 Scheme of laser triangulation [25]

a, b parameters of mapping algorithm possible to be calculated based on known parameters of experimental set-up or which can be gained from calibration of this set-up [31].

Using McLaurin relation development (1.4) and through substitution of $c1 = 1/b$, $c2 = a/b$ the following relation for topographic deviation can be gained [31]:

$$\Delta r = c\Delta u - c2(\Delta u)^2 \tag{1.5}$$

when the relation for calculation of topographic deviation is linear and what makes calculation easier. Calibration of this experimental set-up is carried out as follows: reference plane is located at an appropriate distance onto a micro sliding desk where a light trace is projected (point, linear trace). Shifting this reference plane in the direction of the z-axis and scanning of light trace to the computer and evaluating the result in the relation between Δr (shift of a micro shift) and Δu (number of camera pixels by which the trace center has moved), i.e. between parameters a and b [31].

In the case of inherent measuring the measured object fixed to a micro sliding system with ideally linear light trace projected is situated in the set-up instead of the reference plane. Shifting the measured object in the direction perpendicular to the linear trace with the sufficiently short step of the micro engine and transforming of measuring series result in gaining the topographic distances corresponding with illuminated parts of the surface. Approximation of the surface between these parts leads to the consequent gain of complete topographic information on the measured object [31].

1.9.2 Fourier Profilometry

This incoherent topographic method uses optical structure, i.e. intensity sinusoidal grid. The most suitable source of the grid is currently a data projector which allows fast manipulation of the projected image (change of grid period, its phase shift, etc.) [32–34].

Its principle rests in the evaluation of phase change of sinusoidal grid image by the deformed surface of the measured object through comparison with the image phase of the sinusoidal grid on the reference plane. The phase is achieved by Fourier transformation of these two images and topographic deviation is determined using parameters gained during calibration or calculation [32–34].

The experimental set-up of the methods is very simple. It consists of the sinusoidal grid projector, the measured object, and the camera (Fig. 1.18) [32–34].

The projector projects the sinusoidal grid to a suitably located reference plane and its image is detected by the camera on the computer. Mathematically the image can be expressed by the following relation [35]:

$$g_0(x, y) = l_0(x, y) + [1 + V(x, y)\cos(2\pi f_0 x) + \Phi_0(x, y)] \tag{1.6}$$

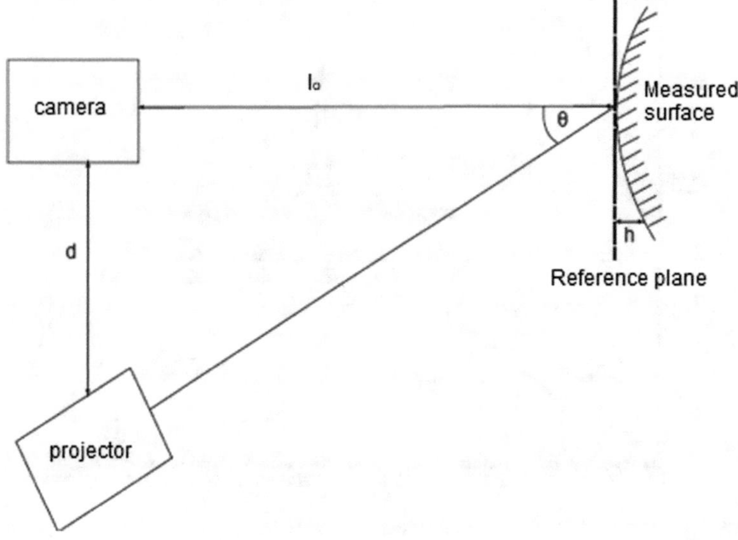

Fig. 1.18 Basic scheme of Fourier profilometry [25]

with

l_0 background intensity,
V stripe visibility,
f_0 grid frequency in the x-direction,
ϕ_0 phase.

Instead of the reference plane, the experimental set-up is then extended by the measured object onto which the identical sinusoidal grid is projected. Its image is also saved on the computer and is described by the following relation [35]:

$$g(x, y) = l_0(x, y) + [1 + V(x, y)\cos(2\pi f_0 x) + \Phi_0(x, y)] \tag{1.7}$$

Consequently, Fourier transformation is carried out in the case of these functions. Fourier function spectra (1.6) and (1.7) shall have the form as follows [35]:

$$G_0(f, y) = \int_{-\infty}^{\alpha} g_0(x, y)\exp(-2\pi i f x)\,\mathrm{d}x \tag{1.8}$$

$$g(f, y) = \int_{-\infty}^{\alpha} g_0(x, y)\exp(-2\pi i f x)\,\mathrm{d}x \tag{1.9}$$

Consequently, filtration of Fourier spectra of these two functions is carried out. Only the spectrum part corresponding with the frequency of f0 is preserved. The

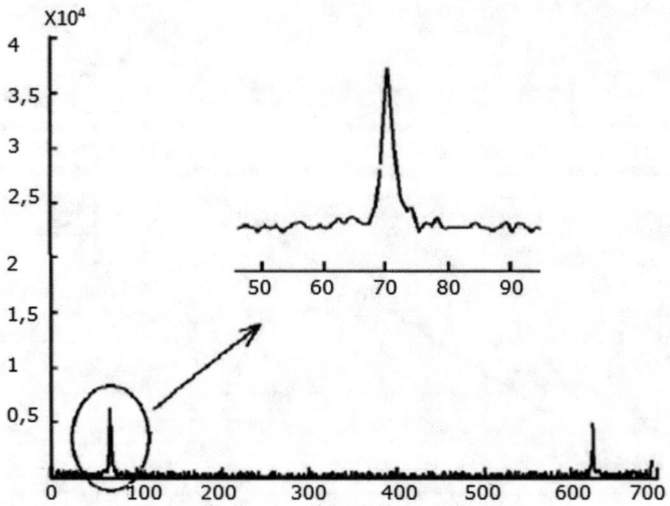

Fig. 1.19 Fourier spectrum of sinusoidal signal with noise [25]

spectrum parts corresponding with other frequencies are set to zero. Figure 1.19 shows the situation in the case in which the Fourier spectrum was calculated from the line of the intensity of the sinusoidal grid image from the reference plane. The x-axis is in pixels and the y axis corresponds with the frequency amplitude [35].

Fourier spectra filtrated as mentioned afore are used in inversion Fourier transformations the results of which are the following functions [35]:

$$g_0^{\wedge}(x, y) = \mathrm{Ar}(x, y)\exp\{i[2\pi f_0 x + \Phi_0(x, y)]\} \tag{1.10}$$

$$g^{\wedge}(x, y) = \mathrm{Ar}(x, y)\exp\{i[2\pi f_0 x + \Phi_0(x, y)]\} \tag{1.11}$$

Equation 1.11 with A standing for amplitude and r standing for amplitude variation. The phase of these complex signals can be expressed by the following equations [35]:

$$\Phi_0(x, y) = \mathrm{arctg}\left[\frac{\mathrm{Im}\left[g_0^{\wedge}(x, y)\right]}{\mathrm{Re}\left[g_0^{\wedge}(x, y)\right]}\right] \tag{1.12}$$

$$\Phi(x, y) = \mathrm{arctg}\left[\frac{\mathrm{Im}\left[g_0^{\wedge}(x, y)\right]}{\mathrm{Re}\left[g_0^{\wedge}(x, y)\right]}\right] \tag{1.13}$$

Then the following relation is applicable for the phase change in the particular point (*x*, *y*) between two aforementioned sinusoidal grids [35]:

$$\Delta\Phi(x, y) = \Phi(x, y) - \Phi_0(x, y) \tag{1.14}$$

There exist two possibilities of transformation of the final matrix of phase change to the matrix of topographic deviations. Such transformation can be carried out either using experimental set-up calibration or through calculation. Direct calculation of phase change to topographic deviation can employ the relation containing parameters of experimental set-up [35]:

$$h(x, y) = \left[\frac{l_0 p_0 \left[\frac{\Delta\Phi(x,y)}{2\pi} \right]}{p_0 \left[\frac{\Delta\Phi(x,y)}{2\pi} \right] - d} \right] \tag{1.15}$$

with

h topographic deviation (Fig. 1.18),
d the distance between camera and projector,
p_0 period of a projected sinusoidal grid divided by the expression cos (θ),
l_0 reference plane distance from monitored plane [35].

The aforementioned process of gaining information on topographic deviation brings about one problem stemming from Eqs. (1.12) and (1.13). The function of arctangent disposes only of the range of values ($-\pi/2$, $\pi/2$) which causes discontinuous throws in the case of phase matrix. Before further adjustment of the matrix, these discontinuities must be eliminated [35].

1.9.3 Phase Shifting Profilometry

It is the case of incoherent optical topographic method which again uses planar light structure, i.e. sinusoidal grid. The data projector is used as the grid projector [36].

The experimental set-up is identical to the one of Fourier PROFILOMETRY (Fig. 1.18). Alike Fourier Profilometry the method is based on the evaluation of phase change of grid images of the reference plane and of the measured object yet they differ in the principle of calculation of phase change. That is not calculated from the argument of complex function gained by inverse Fourier transformation of the filtered Fourier spectrum of sinusoidal grid image but it is calculated from the M sinusoidal grids which are gradually projected to the reference plane and consequently to the measured object and which have a shifted phase by the value of 2π/M. These grids are recorded by the computer and the final phase is then given by the following relation [37].

$$\Phi(x, y) = \left[\frac{\sum_{n-1}^{M} \ln(x, y) \sin(2\pi n/M)}{\sum_{n-1}^{M} \ln(x, y) \cos(2\pi n/M)} \right] \text{arctg} \tag{1.16}$$

with

l_n image intensity of the n-th sinusoidal grid in the particular point.

By calculating the phase of each point of the measured object surface corresponding with the camera chin matrix from the images of the reference plane and of the measured surface, by unpacking their phases and by their deduction the topographic deviation is then calculated [38].

$$\Delta\Phi(x, y) = \Phi(x, y) - \Phi_0(x, y) \tag{1.17}$$

$$h(x, y) = \frac{p.\Delta\Phi(x, y).l_0}{p.\Delta\Phi(x, y) - 2\pi d} \tag{1.18}$$

with

- l_0 a distance of the camera from the reference plane,
- d the distance between camera and projector,
- p wavelength of sinusoidal grid image from the reference plane.

It is worth mentioning that the relation formally corresponds with the relation for the calculation of topographical deviations of Fourier Profilometry (1.15). It becomes apparent that phase-shifting profilometry differs from Fourier Profilometry only in the fashion of phase calculation from the sinusoidal grid image [39].

1.9.4 Phase Measuring Deflectometry

Deflectrometry by phase measuring is in fact modification of phase-shifting profilometry. It is not primarily intended for measuring height profile yet for measuring the measured object curvature [39]. It uses the sinusoidal grid as the light structure but since it is intended for measuring reflection areas the structure is not directly realized by any type of projector, i.e. it is realized by a planar source (Fig. 1.20). In the case of this source, the sinusoidal grid is shown which is recorded by the camera using the measured object as the mirror [40].

The display in the experimental set-up gradually shows M sinusoidal grids which are stepwise shifted about each other by $2\pi/M$ of the phase. The reflections of these sinusoidal grids from the measured object are saved into the computer and on their basis is consequently calculated the phase according to (1.16). Contrary to Fourier Profilometry or phase-shifting profilometry the phase difference of the sinusoidal grid image from the measured object and the reference plane is not calculated but the phase forms a kind of coordinate (Fig. 1.24), i.e. it defines which point of the light source is focused on by the particular camera pixel [41].

According to Fig. 1.21, it is clear that the shift length of the evaluated point on the camera chip matrix from the flat mirror spot on which it was located during calibration allows the calculating magnitude of the measured surface curvature in each point corresponding with the camera chip matrix. The following relation can be derived from Fig. 1.21 for the phase [42]:

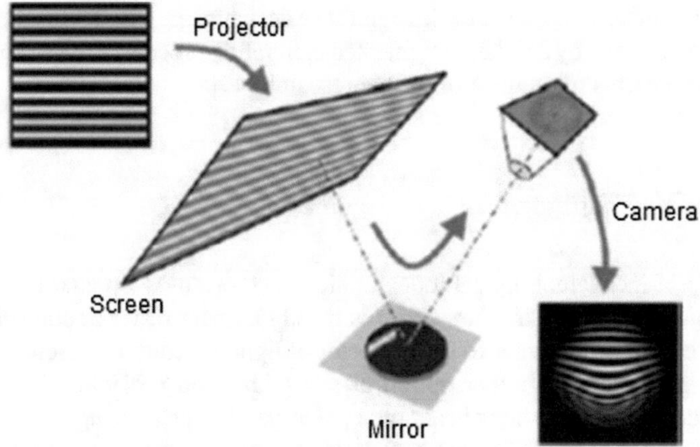

Fig. 1.20 Arrangement of phase measuring deflectometry [25]

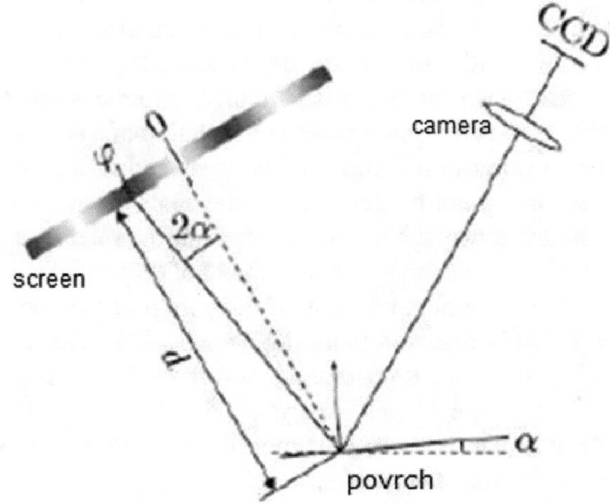

Fig. 1.21 Principle of phase measuring deflectometry [25]

$$\Phi(x, y) = d \cdot \tan 2\alpha(x, y) \tag{1.19}$$

From which by a simple modification the following relation for the searched derivation (surface curvature) in the particular point can be derived:

$$\alpha(x, y) = \frac{\arctan\left[\frac{\Phi(x,y)}{d}\right]}{2} \tag{1.20}$$

In this fashion, all the pixels of the camera chip matrix corresponding with the measured object points are evaluated through which a matrix of primary derivations of the measured surface is formed. This experimental method is used especially in the case of curvature measuring of the lens and spherical mirrors. The development of the method is restrained by a rather complicated calibration procedure the simplification of which represents the subject of further research [43].

1.9.5 *Moiré Topography*

This group of incoherent optical topographical methods uses a planar periodic grid as the projected optical structure and it is mostly a linear raster or sinusoidal grid. Formerly used data projector as the projector of light structure is currently replaced by the data projector. Another option of sinusoidal grid projection is the use of interference of divergent laser beam on a plane-parallel plate using which the light sinusoidal grid is formed. The method itself does not use the interference principle but the superposition principle [42–44].

The principle of these methods is the so-called Moiré phenomenon [45] manifesting itself by the formation of periodical structure from light and dark fringes when two periodical structures of similar parameters overlap each other out of which one is only slightly deformed or slewed about the other. The phenomenon becomes more evident when the original periodical structures are more similar. Overlapping of centers of dark fringes causes the creation of a set-up of moiré fringes. If one of the periodical grids is located on the measured surface and the other is in the reference plane the moiré fringes connect the spots of the measured object surface with the identical topographical deviation from the reference plane, i.e. like in the case of interference methods they form kind of "topographical contours"[45].

Two basic modifications of this group of methods exist—shadow moiré topography and projection moiré topography. The scheme of experimental arrangement of shadow moiré topography is shown in Fig. 1.22 [46, 47].

In the case of this experimental arrangement, the raster is located above the measured surface which defines the reference plane. Raster leakiness is defined by the following relation [45]:

$$T(x, y) = \frac{1}{2}\left\{1 + \mathrm{sgn}\left[\exp\left(i2\pi\frac{x}{p}\right)\right]\right\} \tag{1.21}$$

with

i imaginary unit,
x coordinates,
p raster period.

The light grid on the measured object surface is formed through raster illuminating and it comprises expired elements of the light beam and raster shadows located above the object. This so-called shading grid is deformed by the measured object shape. The shading grid is monitored through the same raster, i.e. it also forms a reference

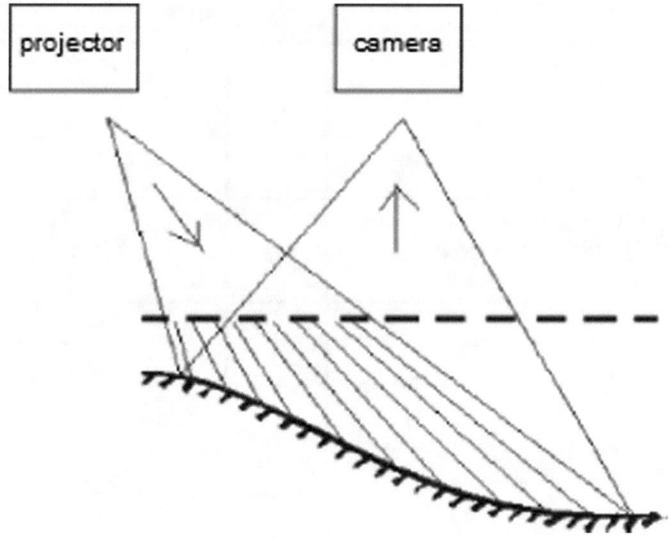

Fig. 1.22 Scheme of shadow moiré topography [25]

grid function. Moiré image consisting of the light grid and the raster is recorded by the camera to the computer and moiré fringes are made of it [45].

Calculation of topographical deviation stems from Fig. 1.23. The y axis is parallel to raster fringes, the x-axis is perpendicular to raster fringes and a linear planar raster is used [45].

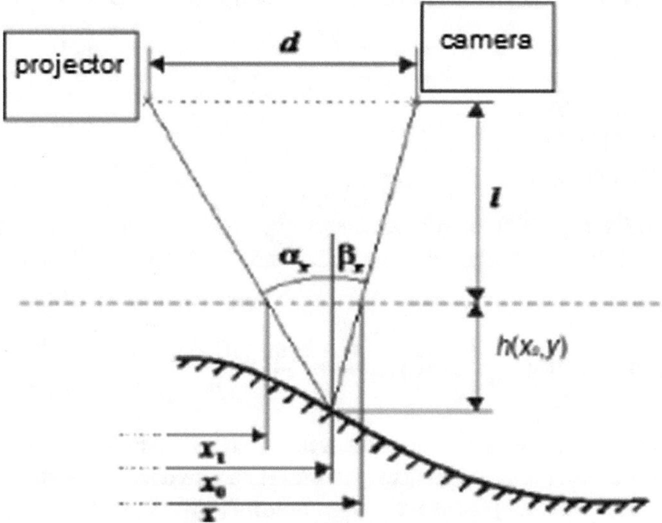

Fig. 1.23 Geometry of shadow moiré method [25]

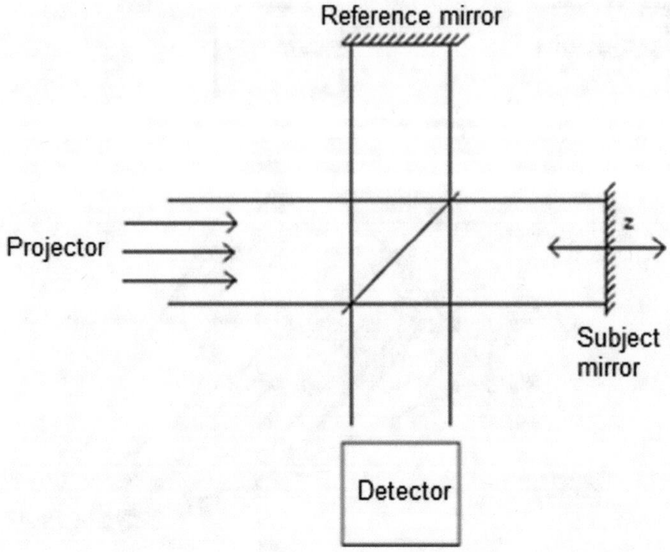

Fig. 1.24 Michelson interferometer [25]

To simplify the calculation it is possible to introduce coordinate transformations according to Fig. 1.23 [45]:

$$x_0 = x_1 + h(x_0, y)tg\,\alpha_x \tag{1.22}$$

$$x = x_0 + h(x_0, y)tg\beta_x = x_1 + h(x_0, y)(tg\alpha_x + tg\beta_x) \tag{1.23}$$

with

h topographic deviation,
α_x projection angle,
β_x observation angle.

Light grid reflected from the measured object surface is in the raster plane described with the use of the relations (1.22) and (1.23) by the relation as follows [46]:

$$I_2(x, y) = \frac{1}{2}\left\{1 + \mathrm{sgn}\left[\exp\left(j2\pi\frac{x - h(x_0, y)(tg\,\alpha_x + tg\,\beta_x)}{p}\right)\right]\right\} \tag{1.24}$$

Equation 1.24 represents the intensity of the moiré grid which shall develop from connecting the raster and the subject grid and it can be derived by substitution of relations (1.21) and (1.24) to the relation as follows [46]:

$$I(x, y) = T(x, y)I_2(x, y) \tag{1.25}$$

$$I(x,y) = \frac{I_0}{4} 1 + \mathrm{sgn}\left[\exp\left\{j2\pi\frac{x}{p}\right\}\right] + \mathrm{sgn}\left[\exp\left\{j2\pi\frac{x_1}{p}\right\}\right] + \mathrm{sgn}\left[\exp\left\{-j2\pi\frac{h(x_0,y)(tg\,\alpha_x + tg\,\beta_x)}{p}\right\}\right] \tag{1.26}$$

In the case of which the second and the third element in the brackets represent formal periodical structures and at the same time the third element is the final moiré image. Topographical deviation can be gained by expressing the third element phase, i.e. [47]:

$$h(x_0,y) = \frac{N(x,y)p}{tg\,\alpha_x + tg\,\beta_x} \tag{1.27}$$

with

N group of moiré fringes.

Following the last relation, it is clear that moiré fringes form topographic coordinates, i.e. they connect the measured surface spots with the identical topographical deviation. The relation (1.27) has been pre-set for diverse modifications of moiré topographical methods [47].

The second basic modification of moiré topography is the so-called projection moiré topography. Contrary to shadow moiré, a physical raster is not used for the formation of the shading grid yet the periodical light structure is formed directly by the projector. Moiré image is thus formed when two light grids overlap [48–51].

The experimental set-up (Fig. 1.23) is firstly extended by a reference plane onto which the light structure is projected (Fig. 1.22). Its image is recorded by the computer and saved in the computer. Consequently, the measured object is located in the set-up instead of the reference plane and again the light grid is projected onto it. Its image deformed by the measured object surface is recorded by the camera and saved in the computer. In the computer, these two light grids overlap by which the moiré image is formed. The image is consequently processed in the same fashion as in the case of shadow moiré topography. The relation used for the calculation of shadow moiré topography is also applied in the case of the calculation of topographical deviation (1.27) [48–51].

1.9.6 White Light Interferometry

This interferometric topographical method uses an intensity light structure. Contrary to "traditional" interference methods this method uses light with large spectral width (referred to as "white" light) the very short coherence length of which is applied in the principle of this method [52, 53]. LEDs or lamps are used as a light source in the majority of cases. The ray of light of these sources of light radiation disposes of sufficiently large spectral width and short coherence length for the use in the case of this method [52, 53].

The principle of this method rests in the interference of two light areas with very short coherence lengths. The ray of light passes through a semi-transparent mirror and one of its parts is then directed toward the reference mirror whereas the second one impinges onto the measured surface. Interference of these two areas results in the occurrence of interference fringes which like in the case of moiré topography have functions of topographical contours, i.e. they connect the spots with the identical topographical deviation (reference mirror has the function of reference plane) [52, 53].

The basis of the experimental set-up of this method is the Michelson interferometer (Fig. 1.24).

Instead of the subject mirrored the measured surface is located in the interferometer which is positioned on the linear micro sliding desk allowing precisely defined movement with a very short (expressed in micrometers) annular step on the z-axis. The camera and the micro sliding desk are connected to the computer. For each point (x, y) of the measured surface, it is applicable that the intensity of the reflected light shall be changing along with the changing coordinate z and it shall reach maximum if its distance from the semi-transparent mirror shall be identical to the distance of the corresponding point on the reference mirror z_0. The dependence of intensity on the position of the slide is shown by the interferogram (Fig. 1.25) [52, 53].

The range of micro slide positions, in the case of which intensity modulation is known, is referred to as interferogram width. The width is directly proportional to the coherence length of the light source and indirectly proportional to its spectral width. To assure the most precise specification of the center of maximal interferogram value, the interferogram must be as short as possible. In the case of the use of a light source with considerable coherence length and narrow spectral width the interferogram would be rather long and within the range given by the micro slide it would be almost constant so it would not be possible to determine its maximal value. Therefore the

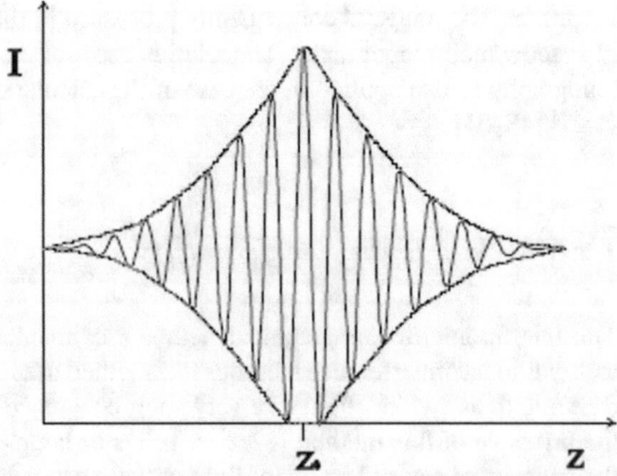

Fig. 1.25 Interferogram—dependence of the intensity of light on the sliding desk position [25]

method uses a light source with short coherence length and extensive spectral width. Dependence of intensity in the measuring point is given by the following expression [54]:

$$I = I_0 \left\{ 1 + \exp\left[-\left(\frac{z}{l_c} \right)^2 \right] \cos\left(4\pi \frac{z - z_0}{\lambda_0} \right) \right\} \tag{1.28}$$

with

l_c coherence light of gauss light,
λ_0 medium wavelength,
I_0 source intensity.

It is the case of the connection between the optical topographical method and the mechanical method. During the experiment, the micro slide moves within the appropriately selected length range from one end to the other by a given number of steps. After each step, it stops and the camera takes a shot of the superposition of two light waves (one from the reference mirror and the other from the measured surface). When the entire measurement process is completed, the interferogram is produced for each point of the measured surface corresponding with the camera chip pixel and its maximum is detected. Thus the coordinate of the z_0 points is formed the height difference of which is given by their size and by several steps made by the micro slide. Using adequate determination of topographical plane position (it can be the lowest value of the group of the z_0 points) and by deduction of the measured coordinates of all points the matrix of topographical deviations is formed [54].

1.9.7 Fizeau Interferometry

Fizeau interferometry (Fig. 1.26) ranks among the groups of coherent optical topographical methods. They use intensity light structure as the optical structure and laser or laser diode as the coherent source [55–57].

Contrary to the majority of other interferometric topographical methods in this case no semi-permeable element is used for dividing the light beam, but the light beam passes through a transparent optical element the back interface (glass-air) of which forms the reference area toward the measured surface (in case of the measured surface a particular level of optical reflectivity must be visible). Part of the light beam is reflected from the measured surface back to the optical element after guarantee and along with the part of the beam reflected from its reference area it interferes. The final interference image is consequently monitored by the camera through the semi-transparent mirror. The basic experimental scheme is shown in Fig. 1.26 [58].

The result of measurement is the interference image comprising of interference fringes with the property of topographic contours—they connect the spots with the identical topographical deviation of the measured object from the reference area [58].

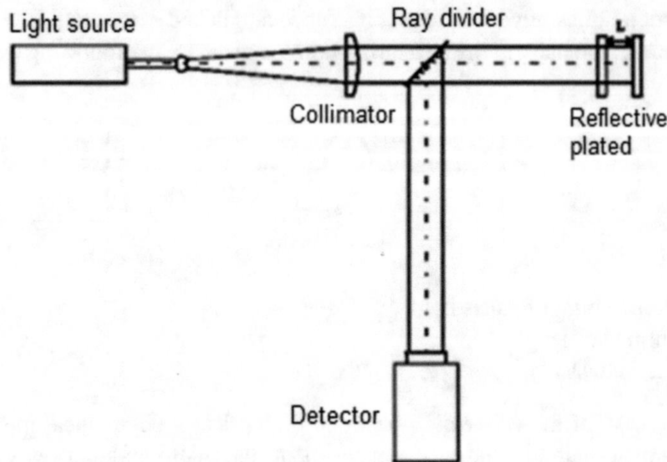

Fig. 1.26 Scheme of Fizeau interferometer

1.10 Comparison of Parameters of Equipment Applied in Surface Roughness Measurement

In case of measurement the contact roughness tester Surftes SJ—400 by the company of Mitutoyo was used as well as laser profilometer LPM. Table 1.12 shows basic quantities of the most frequently used roughness testers [59].

Mitutoyo (Surftest SJ) 400 is designed to measure surface roughness in the case of shop surfaces. The tester comprises of contact stylus which measures the surface structure and evaluates surface quality using a series of parameters by national and international standards [59].

Table 1.12 Comparison of technical parameters and prices of the selected profilometers [59]

Equipment	Mitutoyo SJ—400	LPM
Range z	800 μm	420–470 mm
Measurement range x, y	25 mm	300 mm (slide range)
Processing	0.05-mm/s	2–5000 steps/s
Material samples	metal, ceramics, glass, quartz, plastics	Diverse types of metal materials, plastics, rubber, wood
Measured parameters	R_a, R_z	R_a, R_z, R_q, R_p, R_v

References

1. Liu, J., Lu, E.H., Yi, H.A., Wang, M.H., Ao, P.: A new surface roughness measurement method based on a color distribution statistical matrix. Measurement **103**, 165–178 (2017). https://doi.org/10.1016/j.measurement.2017.02.036
2. Bass, M.: aj. Handbook of Optics. Vol. II. Devices, Measurement and Properties, 2nd edn. McGraw–Hill, New York (1995)
3. Dyson, J.: Interferometry as a Measuring Tool. Machinery Publishing (1970)
4. West, C.M.: Holografická interferometrie. John Wiley & Sons, New York (1979)
5. Hrabovský, M., Bača, Z., Horváth, P.: Koherenční zrnitost v optice. Vydavatelství Univerzity Palackého, Olomouc (2001)
6. Guenther, R.D.: Modern Optics. John Wiley & Sons, New York (1990)
7. Bumbálek, B., Obvody, V., Ošťádal, B.: Drsnost povrchu. Praha: SNTL, 338s (1989)
8. STN EN ISO 4287 –2 Drsnosť povrchu. Terminológia. Časť 2. Meranie parametrov drsnosti povrchu
9. STN EN ISO 4287 Geometrické špecifikácie výrobkov (GPS). Charakter povrchu: Profilová metóda—Termíny, definície a parametre charakteru povrchu (1999)
10. Valíček, J., Rusnák, J., Müller, M., Hrabě, P., Kadnár, M., Hloch, S., Kušnerová, M.: Geometrické aspekty drsnosti povrchu klasických a netradičních technologií. Jemná mechanika a optika, č. 9, s. 249–253 (2008)
11. Tichá, Š.: Strojírenska metrologie Část 2, Základy řízeni jakosti. Ostrava. Vysoká škola baňská – Technická univerzita Ostrava. 86 s. ISBN 8024812096 (2006)
12. Kolnerová, M.: Povrchy povlaků—mikrogeometrie. [online]. Liberec: Technická univerzita v Liberci—Katedra strojírenské technlogie. Oddelení tváření kovů a plastů. 2010. [26.1.2016]. Available on the Internet: http://www.ksp.tul.cz/cz/kpt/obsah/vyuka/stud_materialy/spt/povrchy%20povlaku.pdf
13. Häusler G.: Three-dimensional sensors-potentials and limitations. In: Handbook of Computer Visions and Applications, vol. 1, pp. 485–506, Sensors and Imaging, Academic Press (1999)
14. Fekete, F.: Prístroje pre meranie kvality obrábaných povrchov a schémy zapojenia. Bakalárska práca. Prešov. Fakulta výrobných technológií so sídlom v Prešove (2014)
15. Mahr, spol. s r.o.: Katalóg drsnomerov MarSurf PS1. [online]. Available on the Internet: http://www.mahr.cz/scripts/relocateFile.php?ContentID=5306&NodeID=10778&FileID=11429&ContentDataID=14172&save=0
16. Microtes.: Katalóg drsnomerov CarlZeiss. [online]. Available on the Internet: http://www.microtes.cz/images/Handysurf-E-35B-cz.pdf
17. Dotykové, nebo bezdotykové měření struktury povrchu?: Dotykový způsob měření. MM: průmyslové spektrum. [online]. (2005). Available on the Internet: http://www.mmspektrum.com/clanek/dotykove-nebo-bezdotykove-merenistruktury-povrchu.html
18. Jurena, P.: Snímaní a hodnocení jakosti broušeného povrchu kontaktním a bezkontaktním způsobem: laserový snímač. [online]. Zlín. (2011). Available on the Internet: https://portal.utb.cz/wps/PA_StagPortletsJSR168/KvalifPraceDownloadServlet?typ=1&adipidno=19095
19. Mobilný drsnomer MarSurf. [online]. Available on the Internet: http://www.klz.inshop.cz/6910232-mobilni-drsnomer-marsurf-ps10-snimaci-hrot-5-%C2%B5m
20. For school.eu. Laserová profilometria. [online]. Available on the Internet: http://www.forschool.eu/sk/profilometria/
21. Inspection Equipment: Katalóg drsnomeru MarSurf WS 1. [online]. Available on the Internet: http://www.inspection.ie/shopping_admin/images/prod_img/relocateFileWS.pdf
22. TaylorHobson.: CCI HD. [online]. [27.1.2016]. Available on the Internet: http://www.taylor-hobson.com/products/23/109.html
23. Šustek, J.: Laserový profilometer LPM s horizontálnym posunom pri sledovaní nerovnosti povrchu. *Trieskové a beztrieskové obrábanie dreva 2010*: VII. medzinárodná vedecká konferencia. Zvolen: Technická univerzita vo Zvolene, pp. 187–192 (2010). ISBN 978-80-228-2143-8
24. http://silnylaser.sk/14-informacie/62-farba-a-vykon-laseru

25. M.: Vybrané optické 3D metody a jejich aplikace (2012). Univerzita palackého, Olomouc. https://product.item24.de/en/products/product-catalogue/products/mb-building-kit-for-mechanical-engineeringPOCHMON
26. Menn, M.: Practical Optics. San Diego, Elsevier Academic Press, (2004). ISBN 0-12-490951-5
27. Mitaľ, G., Ružbarský, J.: Bezkontaktné meranie a vyhodnocovanie drsnosti opracovaných povrchov pomocou laserovej profilometrie. In: ARTEP 2016. Košice: TU, 2016 S. 00-1-00-7.—ISBN 978-80-553-2474-6
28. http://ccd.mii.cz
29. Mandát, D., Rössler, T., Hrabovský, M., Gallo J.: Aplikace optických topografických metod v medicíně. Acta Mechanica Slovaka, Košice,s. 327–332 (2006)
30. Asundi, A., Zhou, W.: Mapping algorithm for 360-deg profilometry with time-delayed integration imaging. Opt. Eng. **38**, 339–343 (1999)
31. Nožka, L., Mandát, D., Hrabovský, M.: The 3D optical scanning topography, Theory, and application. Acta Univ, Palacki. Olomouc, Fac. Rep. Nat. (2003–2004), Physica 42–43, 185–194
32. Takeda, M., Mutoh, K.: Fourier Transform profilometry for the automatic measurement of 3-d object shapes. Appl. Opt. **22**(24), 89 (1983)
33. Bone, Donald J.: Fourier fringe analysis: the two-dimensional phase unwrapping problem. Appl. Opt. **30**(25) (1991)
34. Gruber, M., Häusler, G.: Simple, robust and accurate phase-measuring triangulation. Optik **3**, 118–122 (1992)
35. Dursun A.; Ecevít N., Őzder S.: Application of wavelet and Fourier transforms for the determination of phase and three-dimensional profile. In: Proc. of Int. Conf. on Signal Processing, vol. 1, pp. 168–172 (2003)
36. Rastogi, P.; Gorthi, Sai S.: Fringe projection techniques: Whither we are? Opt. Lasers Eng. **48**(2), 133–140 (2010). ISSN: 0143-8166
37. Li, J., Su, H., Su, X.: Tho-frequency grating used in phase-measuring profilometry. Appl. Opt. **36**(1) (1997)
38. Chang, M, Wan, D.: Automated phase-measuring profilometry. Opt. Lase. Eng. 15(2) (1991)
39. Tang, Y., Su, X., Wu, F., Liu, Y.: A novel phase measuring deflectometry for aspheric mirror test. Opt. Express **17**(22), 19778–19784 (2009)
40. Knauer, M., Kaminski, J., Häusler, G.: Phase measuring deflectometry: a new approach to measuring specular free-form surfaces. In: Proceedings of SPIE, Optical Metrology in Production Engineering, vol. 5457, pp. 366–376 (2004)
41. Zhao, W., Su, X., Liu, Y., Zhang, Q.: Testing an aspheric mirror based on phase measuring deflectometry. Opt. En. **48**(10) (2009)
42. Rössler T.: Moderní aspekty a aplikace moiré interferometrie. Disertační práce, PřF UP Olomouc (2003)
43. Post, D., Han, B., Ifju, P.: High Sensitivity Moiré: Experimental Analysis for Mechanics and Materials. Springer-Verlag, New York (1994)
44. Durrell, A., Parks, V.: Moiré Analysis of Strain. Prentice-Hall, Englewood Cliffs, New Jersey (1970)
45. Parks, V.J.: Handbook on Experimental Mechanics. VCH Publishers, New York (1993)
46. Bartl, J., Fira, R., Hain, M.: Inspection of the surface by the Moir´e method, measurement. Sci. Rev. **1**(1) (2001). ISSN 1335-8871
47. Yoshiyawa, T., Tomisawa, T.: Shadow moiré topography using the phaseshift Metod. Opt. Eng. **32**(7), 1668–1674 (1993)
48. Asundi, A.: Moiré methods using computer-generated ratings. Opt. Eng. **32**(1), 107–116 (1993)
49. Keprt, J., Toma, J.: The use of video technique in some applications of moiré topography. Acta UP Ol., Fac. Rer. Nat. Physica **30**, 61–80 (1991)
50. Cchol, Y.-B., Kim, S.-W.: Phase-shifting grating projection moiré topography. Opt. Eng. **37**(2), 1005–1010 (1998)
51. Takasaki, H.: Moiré topography. Proc. ICO Conf. Opt. Methods in Sci. Ind. Meas. 441–446 (1974)

52. Pavlíček, P., Hýbl, O.: White-light interferometry on rough surfaces—measurement uncertainty caused by surface roughness. Appl. Opt. **47**(16), 2941–2949 (2008)
53. Saraç, Z., Groß, R., Richter, C., Wiesner, B., Häusler, G.: Optimization of white light interferometry on rough surfaces based on error analysis. Optik-Int. J. Light Electron Opt. **115**(8), 351–357 (2004)
54. Pavlíček, P.: Měření výškového profilu předmětu pomocí interference v bílém světle. Jemná mechanika a optika, str. 83–85 (2002)
55. Malacara, D.: Optical Shop Testing, 2nd edn. John Wiley & Sons, New York (1992)
56. Reiser, C., Lopert, R.: Laser wavemeter with solid Fizeau wedge interferometer. Appl. Opt. **27**(17), 3656 (1988)
57. Gray, D., et al.: Simple compact Fizeau wavemeter. Appl. Opt. **25**(8), 1339 (1986)
58. Pochmon, M., Rössler, T., Mandát, D., Gallo, J., Hrabovský, M.: Principles of Fizeau interferometer in low-weared ceramic total hip endoprosthesis. In: Popiolek-Masajada, A., Jankowska, E., Urbanczyk, W. (eds.) 16th Polish-Slovak-Czech Optical Conference on Wave and Quantum Aspects of Contemporary Optics, 8–12 September 2008, Polanica Zdroj, Poland. SPIE vol. 7141, pp. 71410Y(1)–71410Y(6) (2008). ISBN 9780819473837. https://doi.org/10.1117/12.822379
59. Gerková, J.: Návrh systému pre meranie, vyhodnocovanie, modelovanie a simuláciu kvality povrchov obrobených technológiou vodného lúča. Dizeračná práca. FVT TUKE Prešov (2015)

Chapter 2
Laser Profilometer Design

Based on existing solutions and available literature three main parts of the contactless system have been designed, i.e. a mechanical part, an optical part, and a control part. The mechanical part comprises a supporting frame consisting of components with vertical adjustment of a measuring head position and a programmable slide of the sample in the *X* and the *Y* axes using a stepper motor. The optical part comprises laser radiation, an objective, and a camera with a CMOS sensor placed in a single localizer. The control part consists of a PC with the utility and the assessment software. The basic method of evaluation of the surface geometrical features is contactless surface profile measurement by laser profilometry (LPM). The system works on the triangulation principle and the measured object is a thin laser line projected at an angle. The reflection of the laser line is recorded by the digital camera. The recorded image serves for the evaluation of the object profile in the cross section determined by the laser line. The laser line projected to the object within the working range of the equipment is generated by a laser modulus (Fig. 2.1).

2.1 Frame Structure

Under the study of diverse types of structures, the most suitable structure appears to be the contactless system the bases of which are precise satin anodized aluminum profiles with longitudinal channels and holes for fixation of connection elements and extensive fittings. The LPM frame structure made of satin anodized aluminum profiles (Fig. 2.2) must comply with the following requirements:

- The simplicity of assembly,
- The simplicity of manipulation,
- Resistance to ambient conditions (scratching, corrosion),
- Low weight,

J. Ružbarský, *Contactless System for Measurement and Evaluation of Machined Surfaces*, SpringerBriefs in Applied Sciences and Technology,
https://doi.org/10.1007/978-3-031-08981-7_2

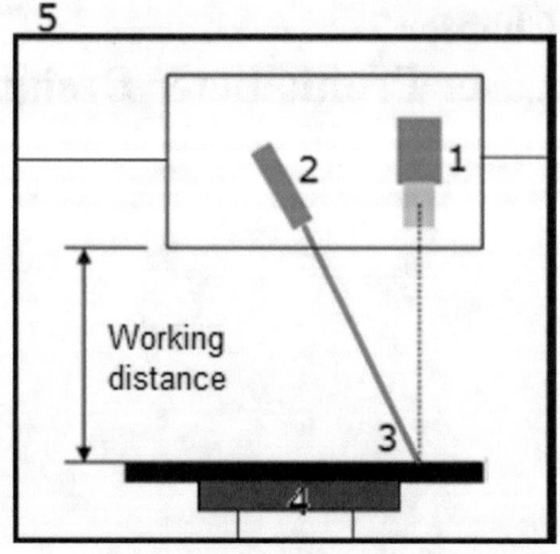

Fig. 2.1 Measurement principle by LPM 1—CMOS Camera, 2—laser light source, 3—measured surface, 4—stepper motors, 5—frame structure

Fig. 2.2 LPM structure profile [1]

- Fast and simple designing,
- Unlimited mobility of the individual profiles about each other without the necessity of further structural frame adjustment,
- High precision, flexibility, strength, and long service life of the individual components with preservation of low weight of the structural unit,
- Affordable solution owing to simple production of manufacture, simple assembly, and the possibility of repeated use of the individual components,
- wide selection of elements—four dimension series of precise satin anodized aluminum profiles with several accessories and fittings.

To assure good stability of the frame parts onto which power elements of laser profilometer are fixed, i.e. its most important section (localizer with laser, objective, and camera) it is inevitable to use corner clasps (Fig. 2.3). The corner clasp consists of a corner body of a triangle shape and two screws with holders inserted into the profile channel.

Fig. 2.3 Corner stabilizing clasp [1]

Dimensions of frame structure depend mainly on the use of other parts of the system such as stepper motors of sliding and localizer with laser, objective, and camera. The proposed dimensions of the frame structure are 730 × 660 × 660mm.

2.2 Micrometric Slide

Based on a study of systems with slides in two axes a micro-metric horizontal slide in the axes *x*, *y* was designed which is assured by two stepper motors Standa 8MT160-300 (Fig. 2.4). The slide in the axis *z* is solved by the mechanical lift of the entire profilometric head using block bolts. The stepper motor must assure precise switching of trajectory to 300 mm and motorized switching with a high definition of 2.5 μm per step or 0.31 μm per 1/8 step. Motorized switching should comply with the requirement of manual control which should be carried out with a rotating button with accessories fixed to the end of the motor structure. These motors were designed stemming from the fact that their basic features include high stability, compact and monolithic design, and high definition which make the machine ideal for integration

Fig. 2.4 Stepper motors Standa 8MT160-300—*x*, *y* [2]

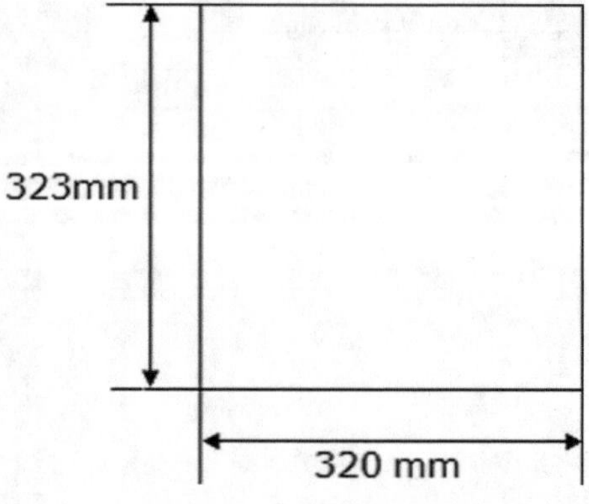

Fig. 2.5 Set-up of slides and work desk with the measured sample

Table 2.1 Specifications of the stepper motor regulator

Definition	Full step, 1/2 of step, 1/4 of step, 1/8 of step
Speed	2–5000 steps s^{-1}
Programmable acceleration and deceleration	–
Two programmable end switches	for each axis
Control using interface	RS232
Max. operating temperature	70 °C
Dimensions	170 × 175 × 85 mm

into highly precise measurement systems. Slide carrying capacity is of 10 kg; definition in the x-axis is of 22/1230 = 0.018 mm/pix and in the x-axis it reaches 7/340 = 0.021 mm/pix. Maximal slide speed amounts to 8 mm s^{-1}.

Part of the slide is a work desk attached to the stepper motor onto which a measured sample is fixed during the measuring. Dimensions of the work desk should be 320 × 323 mm, which means that the maximal dimension of the measured sample can reach 300 × 300 mm (Fig. 2.5).

Part of the stepper motors is a unit for control of motor steps (regulator) which is supplied along with motors. Motors are supplied with the regulator 8MSC1-USBhF-B2 Stepper motor controller. Table 2.1 presents detailed specifications of the motor regulator.

2.3 Objective

Based on the research of objectives for highly precise definition inevitable for this type of measurement, the objective Tamron 23FM50SP is proposed. The objective uses advanced optical techniques for reaching high definition with zero distortion.

Using the latest optical design the objective preserves excellent performance with a minimum focusing distance of only 0.20 m and it is small enough to assure compatibility with current automatized machines. With regards to sharp image clarity in the minimally focused distance, the objective proves to be ideal for any application requiring the recording of shots with high definition which is inevitable for contactless equipment designed for the measurement of surface characteristics (Fig. 2.6).

Table 2.2 shows the specifications of the proposed objective.

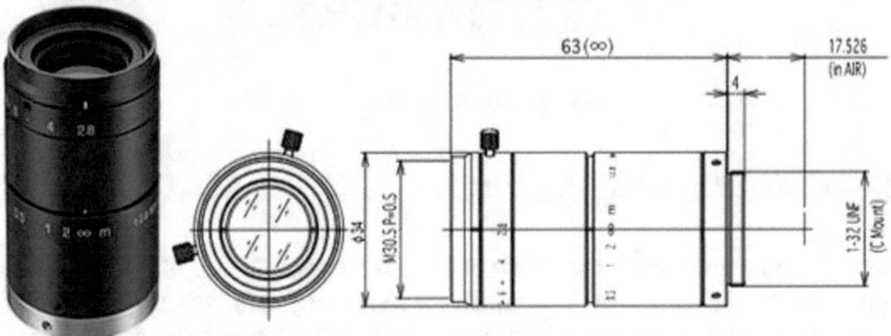

Fig. 2.6 Objective Tamron 23FM50SP 50 mm [3]

Table 2.2 Specifications of objective Tamron 23FM50SP 50 mm

Name	Unit
Size of image	2/3
Type of fixation	C
Focal distance	f = 50 mm
Range of aperture	2.8–32
The angle of view (Horizontally × vertically)	2/3–10.1 × 7.6
	1/2–7.3 × 5.5
	1/3 5.5 × 4.1
TV distortion	Less than 0.01%
Focus range	0.2 m–∞
Working distance	185 mm
Filter size	M30.5 $P = 0.5$ mm
Weight	117 g
Operating temperature	10–60 °C

Fig. 2.7 Digital CMOS camera AVT MARLIN F-131 B [4]

2.4 CMOS Camera

Based on our requirements related to quality and speed of shooting the most suitable appears to be the camera AVT MARLIN F-131B (Fig. 2.7) because it is a rather compact camera with progressive scanning CMOS fitting for the production of highly sophisticated images. MARLIN F-131 B can reach the speed of up to 25 shots per second with the full definition. It is equipped with verified and easy to be used interface FireWire (IEEE 1394a) for high-performance projection in complicated applications.

Vast possibilities of image processing result in high quality of image, reduction of retouching, a lower load of CPU, and higher performance. It can be easily integrated into existing applications owing to a highly efficient and flexible interface API. It is especially suitable for use in the area of industrial processing of images and automatization of products. Table 2.3 presents the specifications of camera AVT MARLIN F-131B.

2.5 Applied Laser

Based on a survey of affordable and suitable lasers the laser StingRay-445Nm (Fig. 2.8) complies with the laser requirements. It is the laser modulus with the highest performance. This laser is used mainly for its precision, resistance, and possibility of the configuration of wavelength and performance.

The laser must meet other requirements such as the following:

- Protection against overheating and extreme turnover,
- Analog or digital modulation,
- Microprocessing control,
- Installation under unfavorable conditions such as high temperature, vibrations, dust, or continuous operation.

Table 2.3 Specifications of CMOS camera AVT MARLIN F-131B

Name	Unit
Sensor	Cypress IBIS5B
Sensor type	Sensor CMOS of 2/3 type (diameter of 11 mm)
Size of image	pixel
Size of cells	6.7 μm × 6.7 μm
Depth of definition	10 bit (ADC)
Digital interface	IEEE 1394 IIDC v. 1.3
Transfer speed	100 Mbit/s, 200 Mbit/s, 400 Mbit/s
Radiofrequency	25 Hz
Manual control	0–16 dB (13 × 1.25 dB)
Speed of shutter	108.864 μs (67 s); automatic shutter
Intelligent functions of correction DSNU	Defect rectification; correction of shading during the actual time
Connection	DC 8–36 V through cable IEEE 1394 or 12-pin HIROSE
Energy consumption	Less than 3 watts (12 V DC)
Dimensions	72 mm × 44 mm × 29 mm
Weight	120 g
Operating temperature	5–45 °C

Fig. 2.8 Blue laser StingRay-445Nm [5]

Table 2.4 Specifications of blue laser StingRay-445 Nm

Name	Unit
Wavelength	445 Nm
Connection	150 mW
Possibility of control	RS-232
Range of supply voltage	5 up to 24 V

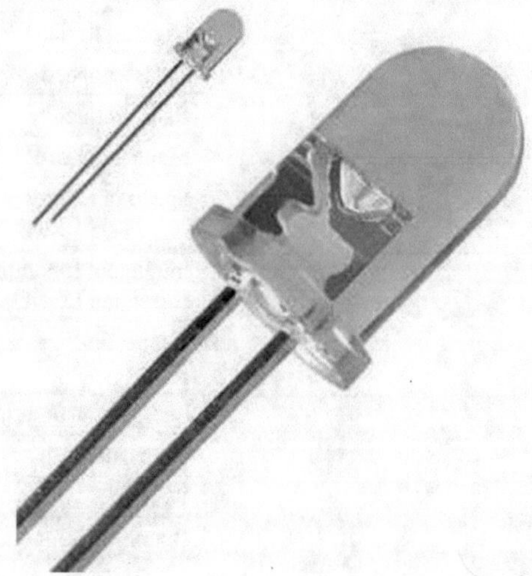

Fig. 2.9 White LED lighting of work area LPM

Specification of the proposed laser StingRay is shown in Table 2.4.

2.6 Lighting

Lighting represents the part of the profilometric head (localizer) of the system designed for contactless characterization of surface profiles. It consists of four white LED lights (Fig. 2.9). It serves as auxiliary lighting and as the lighting of the work desk LPM.

2.7 LPM View Software

Software equipment of the designed system consists of base and utility programs. Base programs include operating systems MS Windows and MS Office (for data export to Excel). Utility programs include the designed software LPM View serving

Table 2.5 Final parameters according to STN EN ISO 4287 and STN EN ISO 11562 standards

Waviness	Complete name of roughness parameter	Roughness
R_p	Height of the maximum peak of the measured profile	W_p
R_v	Depth of the largest valley of the measured profile	W_v
R_z	Maximum height of irregularities of the measured profile	W_z
R_a	Mean arithmetic deviation of the measured profile	W_a
R_q	Mean quadratic deviation of the measured profile	W_q

for communication between LPM and PC and for the setting of measurement parameters of the LPM equipment the part of which is represented by the auxiliary program AVT SmartView used for camera view and creation of image documentation and slide test (slide control and testing). LPM view software is solved within the framework of sub-delivery.

- Through the communication with hardware parts of the equipment the LPM view software should allow the following:
- Range,
- Creation of a 3D model of the measured object profile,
- Change setting of optical system parameters—sample size, measurement density, speed of image processing (measurement)—in dependence on parameters set in the program.
- Export of results to MS Excel—to save the scanned profile into the CSV format,
- Live view—to display an image recorded by a camera located in the LPM equipment,
- Export of images—to save the image from the camera as a bitmap image.

Table 2.5 presents roughness parameters that are measured and evaluated by the designed system.

Roughness parameters Ra (Mean arithmetic deviation of the assessed profile) and Rz (Maximum height of irregularities of the assessed profile) were selected for the experiment because the evaluation of these roughness parameters provides us with a sufficient amount of information.

2.8 General Characteristics of the LPM System

The designed components were used for laser profilometer structure which is intended for contactless characterization of surface geometry (Fig. 2.10). The system allows to measure the samples with a maximum weight of 10 kg with the precision of position setting of 2.5 μm per step and each step consists of 8 μm. The scanner definition is 0.02 mm/pixel.

Using the experimental system it is possible to measure the parameters of surface roughness of the samples according to STN EN ISO 11562 standard (R_q, R_v, R_z,

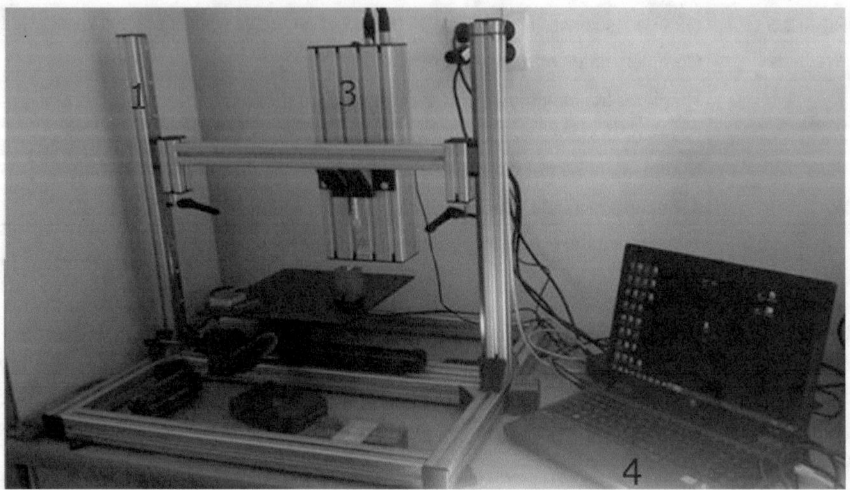

Fig. 2.10 Set-up of laser profilometer LPM 1—the structure of the designed LPM, 2—automated slide of sample in the *X* and *Y* axes, 3—optical part of the system, 4—PC with utility and evaluation software LPM view

R_a, R_p,). Roughness parameters R_a and R_z were selected for the experiment because these parameters provide us with a sufficient amount of information on the quality of the examined surface [6].

The system also includes software LPM view used for LPM control and the evaluation of measured data of the system was solved within the frame of the sub delivery and should comply with requirements of control simplicity and compatibility with a standard PC. Figure 2.11 shows the main window of the software LPM view and consists of the following parameters:

- Menu,
- Toolbar,
- Control panel,
- Adjustment panel,
- Camera window,
- Profile window,
- Status line.

Supported formats are CSV (Comma-Separated Values)—which exports data according to specifications RFC 4180 and XLS. (Microsoft Excel)—it starts the program Microsoft Excel and exports data to a lately opened window of Microsoft Excel.

The user interface is adjustable, i.e. the individual windows can be replaced as required. The control panel serves for the adjustment of parameters of LPM equipment. It consists of three independent tabs. If no other device is connected, only the Device tab is displayed which serves for the selection of devices desired to work

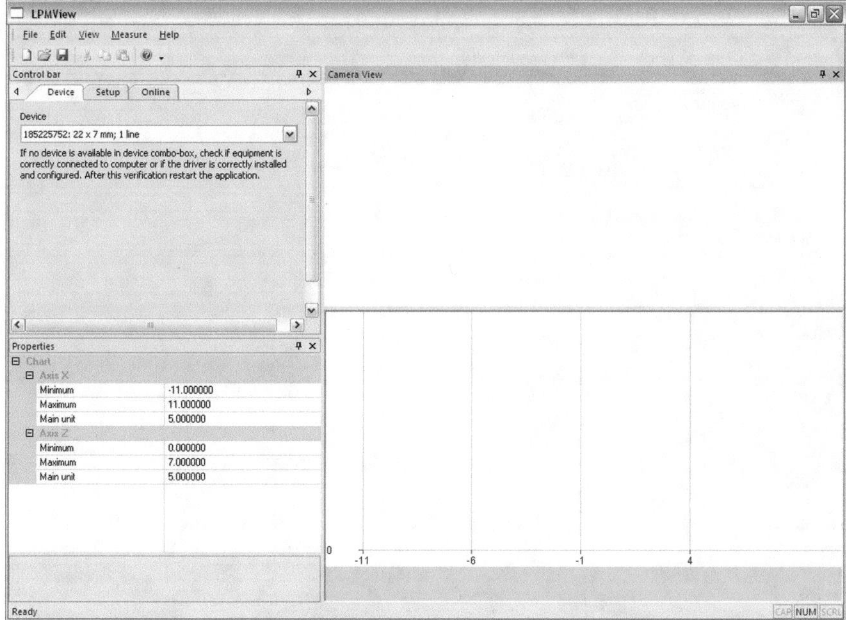

Fig. 2.11 Main screen of the program

with. The item "device" shows available devices (Fig. 2.12). The required device must be selected from the available list.

The set-up card serves for the adjustment of the LPM system (Fig. 2.13). In general, in the case of this dialogue window, it applies that the values entered into the window should fall into the range specified for the particular equipment. The change of the values in this dialogue window can accelerate and fit the measurement for a particular material. The dialogue window does not allow to enter the values out of the operating range of control.

The most significant parameters preset before measurements are time and gain mode. Shutter time sets the exposition time for the camera. The exposition time is

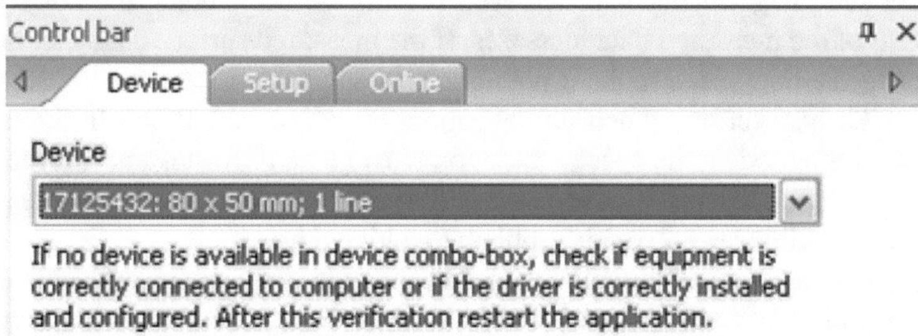

Fig. 2.12 The device tab

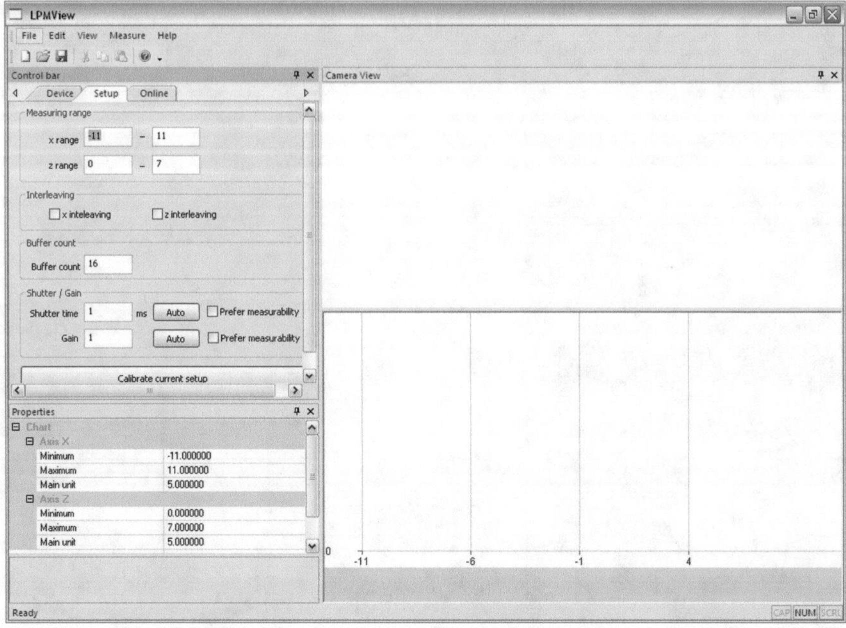

Fig. 2.13 The set-up tab

given in milliseconds. The exposition period defines the time during which the image is scanned. It means that the longer the exposition period is, the more "blurred" the measured data are. On the other hand, to assure visibility and measurability of the laser line it is required to assure also certain exposition period which depends on measured surface properties and the gain mode. After the start-up of the equipment, the exposition period of 1 ms is set. When selecting the Auto mode the automatic fine-tuning of the exposition period is initialized. Before its call, it is inevitable to set the required gain mode. The function attempts to achieve a balance between samples that cannot be measured due to extremely low glossiness and the samples which are slightly noised due to extremely high glossiness. The gain mode (operating mode) sets the internal boosting of the image signal in the camera. The higher it is, the better the image intensity is achieved yet, on the other hand, the noise in the image and of the measured data increases. If the measured surface character and the required measuring speed make it possible, it is advisable to use the lowest value, i.e.1. Implicitly, after equipment initialization, the value of 1 is set. In the case of setting the high mode gain, special attention must be paid (for instance, if the value exceeds 7). Should that be the case, some of the samples can be incorrectly measured. Camera calibration can solve this problem to a high degree.

To display the image of the area scanned by the profilometer camera the LPM view was used (Fig. 2.14) which apart from the view of the scanned area and production of the individual images allows for recording a video of the measured component surface in a combination with the activated slide. To assure effective scanning of the

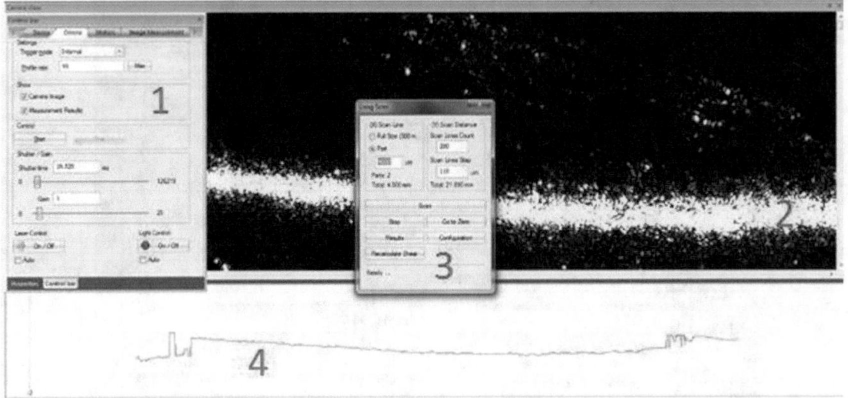

Fig. 2.14 Software LPM View in experimental sample measuring

sample the LPM system is equipped also with an integrated light consisting of four white LED lights. The lighting of the sample by the laser ray or by the LED light is selected in the software part of the system before the surface measuring.

1. The setting up window for parameters of measurement by laser profilometer. In this window, a live camera mode is activated and the operating mode of the profilometer (Gain) along with the time of sample scanning (Shutter time) is set. Correctness of parameter setting determines measured data quality,
2. The window of the actual laser image on the measured sample surface (Camera view). It serves for displaying the image which is provided by the equipment in the mode of the Camera image. The image can be zoomed in and zoomed out and shifted.
3. By clicking on the roughness and waviness window the Long scan window is opened. In this window the parameters of the size of the measured part of the surface are set, i.e. width, length, length of steps of stepper motors in micrometers. This window allows initializing of the measured surface scanning that makes visible in which phase the measuring of roughness parameters occurs.
4. The surface profile window. This window displays the last measured profile in the mode of Measurement results. When the cursor stops on the actual position, the actual value of the cursor position appears. The values are expressed in millimeters.

Fig. 2.15 shows the window with the measured values of the surface roughness which were automatically generated by the LPM software. This window allows the export of the measured data to the formats mentioned herein and to draw a graph of the measured surface profile as well as generate the 3D model of the measured sample (Fig. 2.16). The 3D model can also be compiled from the exported profiles in specialized programs operating with space graphs such as MS Excel or Microcal Origin.

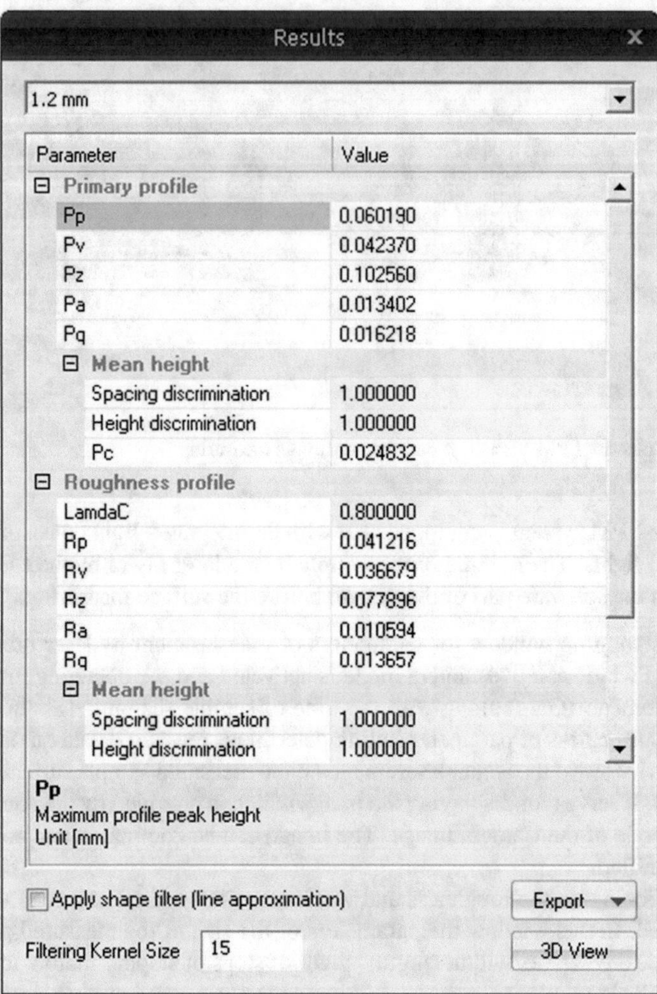

Fig. 2.15 Dialogue window with the measured parameters of roughness and waviness of the surface profile

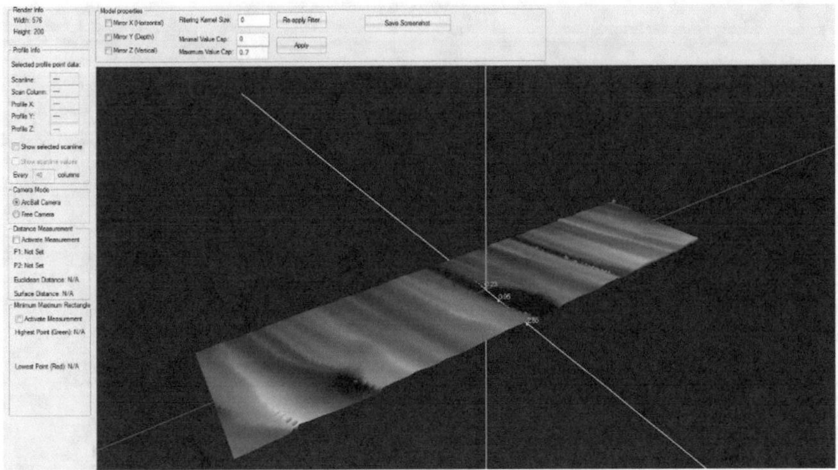

Fig. 2.16 Final 3D model of surface profile

References

1. M.: Vybrané optické 3D metody a jejich aplikace (2012). Univerzita palackého, Olomouc. https://product.item24.de/en/products/product-catalogue/products/mb-building-kit-for-mechanical-engineeringPOCHMON
2. https://www.google.sk/search?q=Standa+8MT160-300&tbm=isch&source=iu&ictx=1&fir=rNDRNgL7JTJbLM%253A%252C1SNIWSTUPMaEaM%252C_&usg=__Drqtq26KqtkczrLtAlyh3fVqoy0%3D&sa=X&ved=0ahUKEwiOoKa07bfbAhUCbFAKHa2qDTMQ9QEIMTAB#imgrc=rNDRNgL7JTJbLM
3. http://www.tamron.biz/en/data/ipcctv/cctv_mg/23fm50sp.html
4. http://www.adept.net.au/cameras/avt/marlin_f131.shtml
5. http://lasers.coherent.com/lasers/660-Nm-Cw-Diode-Laser
6. STN EN ISO 4287 Geometrické špecifikácie výrobkov (GPS). Charakter povrchu: Profilová metóda—Termíny, definície a parametre charakteru povrchu (1999)

Chapter 3
Production of Experimental Samples

The individual experimental samples used for LPM testing are produced by a company that deals with material cutting by abrasive waterjet cutting (AWJ). The samples are made of three material types—aluminum with standardized marking EN5083, stainless steel with standardized marking A304 (stainless steel), and constructional steel with standardized marking S235JR. These materials were selected for being the most frequently used and available materials. Moreover, these materials are applied in practice in manufacturing plants dealing with the issue of material cutting. Each material is used for the production of three samples with identical cutting parameters apart from slide speed which changes. The measured samples produced by the AWJ technology dispose of the same dimensions (Fig. 3.1).

3.1 Material Characteristics of Proposed Samples

Aluminum alloy with standardized marking EN AW-5083 (AlMg 4.5 Mn) is an alloy not possible to be thermally hardened. It is used in the manufacturing of devices and machines (pressure vessels, cryogenic utilization), in the chemical industry (storage tanks), and means of transport. Corrosion resistance is excellent under standard atmospheric conditions or slightly more aggressive conditions such as the marine environment. This material can be sensitive to intercrystalline corrosion especially after thermal treatment or after long-term use at a temperature of >65 °C.

Stainless steel was characterized by the standard as follows: STN 17 240, 17 241 W Nr. 1.4301 A 304. It is the case of austenitic chrome-nickel steel which is the most commonly used type of stainless material with high resistance to corrosion, cold workability, and weldability. It is resistant to water, water vapor, air moisture, edible acids, and weak organic and inorganic acids.

Constructional steel was characterized by the standard STN 11 373 (S235JR). It is the case of unalloyed steel with standard properties which is suitable for welding

J. Ružbarský, *Contactless System for Measurement and Evaluation of Machined Surfaces*, SpringerBriefs in Applied Sciences and Technology,
https://doi.org/10.1007/978-3-031-08981-7_3

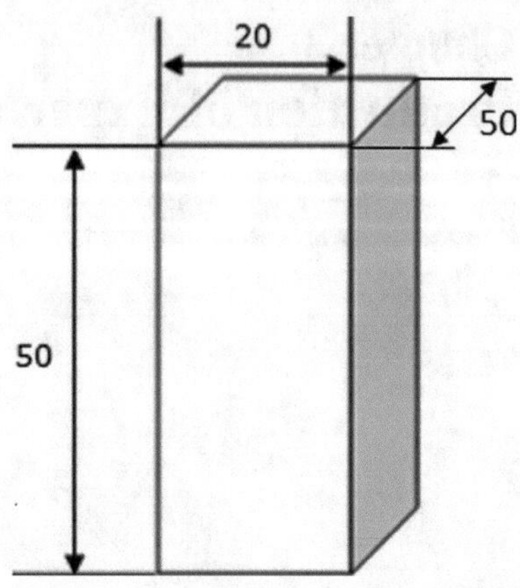

Fig. 3.1 Dimensional specification of experimental samples produced by the AWJ technology

steel structures. At the same time, this unalloyed steel is used as part of low thickness constructions and machines which are fusion welded and statically and slightly dynamically stressed.

3.2 Production of Samples by AWJ Technology

The samples which were used for measuring using the LPM system were produced by material cutting with the abrasive waterjet cutting (AWJ). Table 3.1 presents parameters of cutting machine adjustment in cutting of samples of stainless and constructional steel and aluminum by the AWJ technology which are constant (apart from cutting head speed) in case of all produced samples. Figure 3.2 shows a workstation of the water jet. It is a coordinate table WJ 3020b-1Z by the PTV Praha company and a high-pressure pump 9XD 55 by the FLOW SYSTEMS company. The experimental samples were produced with the use of abrasive material—Australian Garnet with a grain size of MESH 80. Speed of cutting head slide represents one of the most important and technologically one of the simplest controllable technological parameters about quality parameters Ra and Rz [1]. This fact is also proved by the measuring experiment of produced samples.

Table 3.1 Machine input parameters in case of abrasive waterjet cutting

Parameter	Value
Pressure	370 (MPa)
Abrasive type	Australian Garnet
Abrasive grain size	MESH 80
Water nozzle diameter	0.406 (mm)
Rectifying tube diameter	0.889 (mm)
Amount of abrasive	430 (g min^{-1})

Fig. 3.2 Coordinate table WJ 3020b-1Z for the AWJ technology

Machine parameters:

- This configuration can cut metal materials of thickness reaching even 200 mm,
- Effective working zone is 2000 mm × 3000 mm,
- Stroke of 200 mm.

Figure 3.3 shows the material samples which were cut by the abrasive waterjet technology. Figure 3.3 shows the type of sample material, speed of cutting head in case of a particular sample, and marking (nomenclature) of the individual samples.

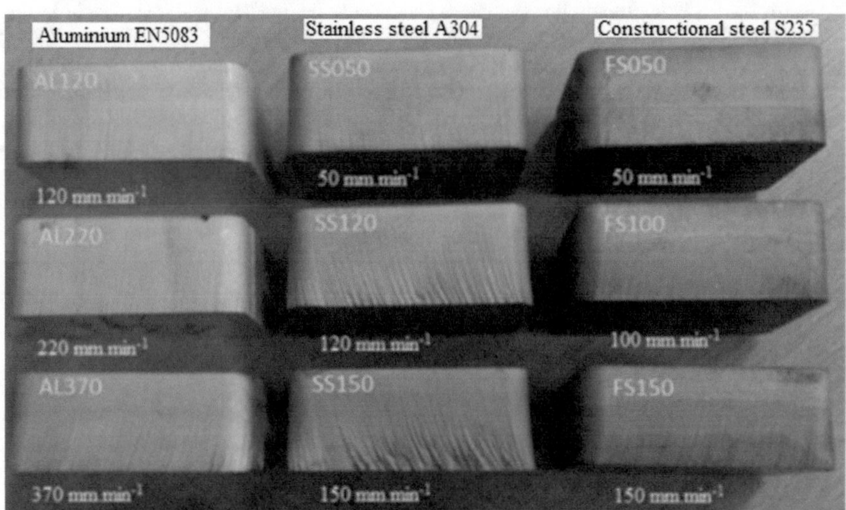

Fig. 3.3 Produced experimental samples

Table 3.2 Roughness declared by the company

Marking of aluminum samples	AL120	AL220	AL370
Declared roughness	6.3 μm	12.5 μm	25 μm
Marking of stainless steel samples	SS050	SS120	SS150
Declared roughness	25 μm	12.5 μm	6.3 μm
Marking of constructional steel samples	FS050	FS100	FS150
Declared roughness	25 μm	12.5 μm	6.3 μm

Material cutting was in charge of the company which in case of a particular material and slide speed of cutting head during cutting declares (about its customers) certain surface roughness. Verification of values of surface roughness of prepared samples independence on the cutting head slide speed of the machine using the focused laser profilometer represents along with a comparison of the values measured by the contact method a significant part of the experimental section of the dissertation. The values of declared surface roughness in the case of different slide speeds of the cutting head are shown in Table 3.2. The company declares that the samples produced by the AWJ technology should not overlap the roughness in any part of the surface. The work written by Lu [2] and by Kang and Liu [3] proves that workpiece surface roughness is the most important characteristic of product quality. Achieving the desired surface quality is rather significant for the functional behavior of components.

References

1. Gerková, J.: Návrh systému pre meranie, vyhodnocovanie, modelovanie a simuláciu kvality povrchov obrobených technológiou vodného lúča. Dizeračná práca. FVT TUKE Prešov (2015)
2. Kang, C., Liu, H.X.: Small-scale morphological features on a solid surface processed by high-pressure abrasive water jet. Materials **6**(8), 3514–3529 (2013). https://doi.org/10.3390/ma6083514
3. Lu, C.: Study on prediction of surface quality in the machining process. J. Mater. Process. Technol. **205**(1–3), 439–450 (2008). https://doi.org/10.1016/j.jmatprotec.2007.11.270

Chapter 4
Description of Measuring Experiment

A high-speed waterjet creates a relief in the workpiece which can be divided according to the surface character into several zones—smooth, medium rough, rough. The task of the experiment is the evaluation and comparison of the measured data of the surface of samples produced by the AWJ technology in the factory. The monitored parameters are R_a—arithmetical mean deviation of the profile (Fig. 1.3) and R_z—maximal height of irregularity of the assessed profile (Fig. 1.4). In practice, the parameters R_a and R_z are the most frequently monitored roughness parameters and can provide a sufficient amount of information on measured surface quality. The measuring experiment of samples cut by the AWJ technology is carried out with the contactless laser profilometer LPM. Verification of the correctness of the measured values and comparison of the measured values is carried out with the contact roughness tester Mitutoyo SJ 400 (Fig. 4.1).

More detailed specifications of the contact roughness tester are shown in Table 4.1.

4.1 Measurement of Selected Roughness Parameters by Contact Method

The quality values of the machined surface of samples produced by the AWJ technology are in the case of the contact method measured in four steps of depth with each sample (Fig. 4.2). Stepping of measurement by contact roughness tester Mitutoyo SJ 400 starts on the surface in a smooth zone of the sample surface and finishes in a rough zone of the sample surface. In the case of this method, the number of measurements is reduced due to measuring complications and difficulties.

J. Ružbarský, *Contactless System for Measurement and Evaluation of Machined Surfaces*, SpringerBriefs in Applied Sciences and Technology,
https://doi.org/10.1007/978-3-031-08981-7_4

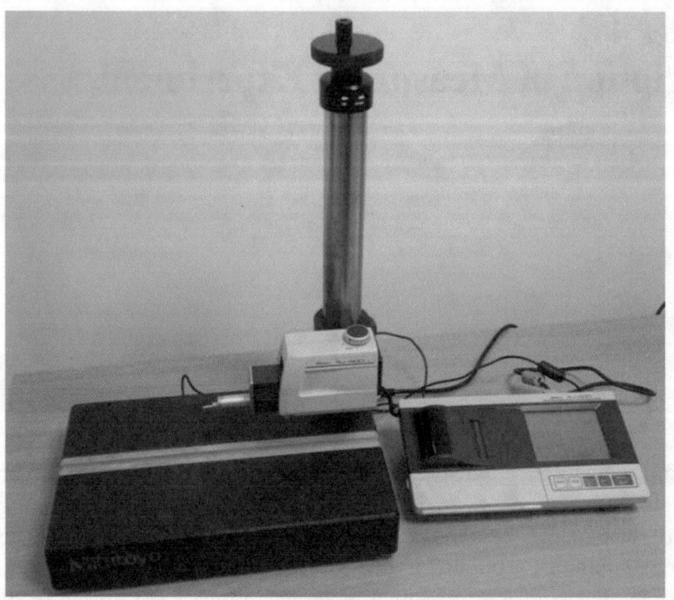

Fig. 4.1 Mitutoyo SJ 400

Table 4.1 Technical parameters of roughness tester Mitutoyo SJ400

Measuring speed	0.05; 0.1; 0.5; 1.0 mm s^{-1}
Speed of return	0.5; 1.0; 2.0 mm s^{-1}
Backward	Backward
Positioning	± 1.5° (angle), 10 mm (upwards/downwards)
Range/definition of measuring	800/0.01 μm; 80/0.001 μm
Type of connection	Using power adapter
Evaluated parameters	*P* (primary), *R* (roughness), W (filtered waviness)
Digital filter	2CR, PC75, Gauss
Cut-off length	0.08; 0.25; 1.8; 2.5; 8 mm

4.2 Measurement of Selected Roughness Parameters by Contactless Method

The values of the machined surface of samples produced by the AWJ technology are measured by laser profilometer in ten steps of the machined area depth according to the scheme shown in Fig. 4.3 with the step of 2 mm. Stepping off the measurement by contactless laser profilometer LPM begins on the surface along the smooth zone of the sample surface and finishes along the rough zone of the sample surface.

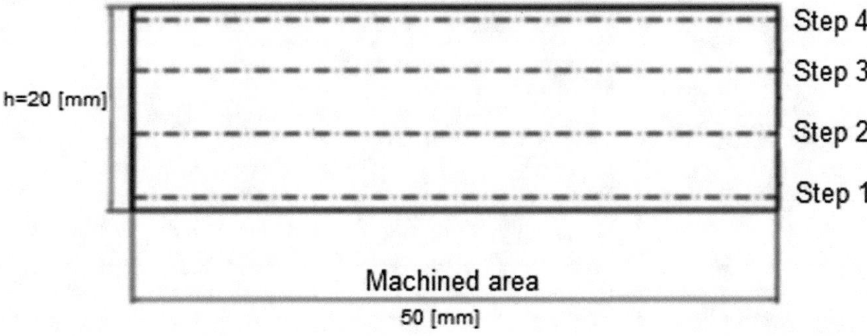

Fig. 4.2 Spots measured by contact method

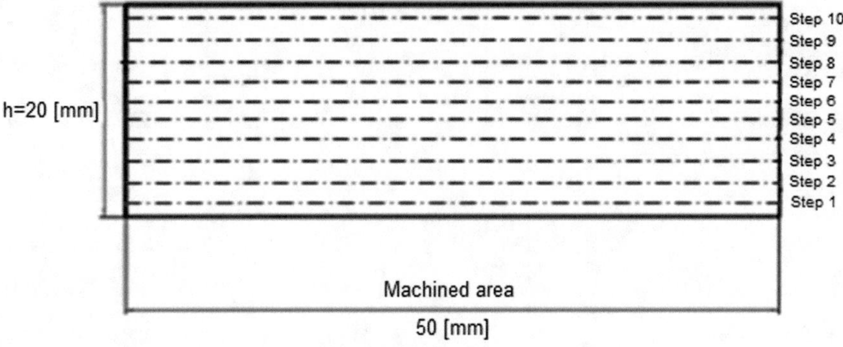

Fig. 4.3 Spots measured by contactless method

Chapter 5
Measurement of Roughness Parameters by Contact Surface Roughness Tester

Measurement of experimental samples produced by the AWJ technology was carried out using the contact method due to verification of the declared roughness of the produced samples and due to comparison of the values measured by contactless laser profilometer LPM with the values measured by contact roughness tester Mitutoyo SJ 400. Measuring was carried out in four steps along the sample surface. The comparison serves for the detection of the correctness of data measurement and evaluation of data measured by the contactless laser profilometer.

Figure 5.1 shows the development of measurement by roughness tester Mitutoyo SJ 400 in case of which in part "a" of the figure the measurement is in progress and the profile being measured is presented by drawing the stylus along the sample surface. In part "b" the measurement touch probe returns to its original position along the same trajectory and evaluates the measured profile. Part "c" shows the evaluated roughness parameters Ra, Rz of the measured surface.

During measurement by contact roughness tester Mitutoyo SJ 400 the experimental samples were inserted into an anti-vibrating substance to eliminate the influence of vibrations of the ambient conditions (Fig. 5.2). A white arrow in the figure shows the direction of movement of the measurement touch probe.

5.1 Samples with Declared Roughness of $R_a = 6.3$ µm (AL120, SS050, FS050)

The measured roughness parameters R_a and R_z of the samples with declared roughness of 6.3 µm are shown in Figs. 5.3 and 5.4.

Figure 5.3 shows the graphical dependence of R_a on stepping for samples with declared roughness of 6.3 µm. The curves from step 2 to step 14 show a linear increase of values of roughness R_a. From step 14 a slight decline in roughness values can be observed in the case of samples AL120 and FS050. In the case of sample

J. Ružbarský, *Contactless System for Measurement and Evaluation of Machined Surfaces*, SpringerBriefs in Applied Sciences and Technology,
https://doi.org/10.1007/978-3-031-08981-7_5

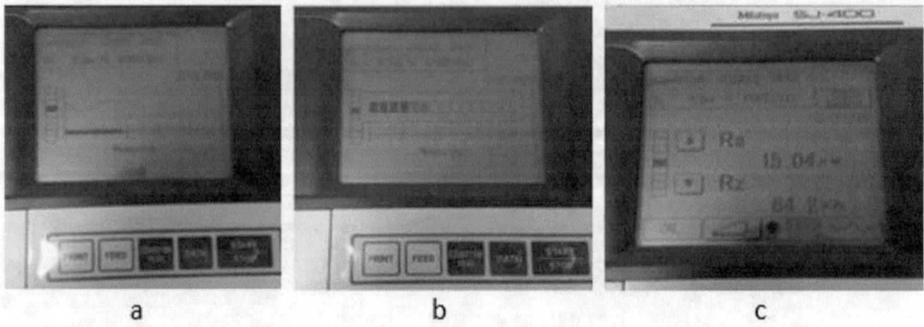

Fig. 5.1 Software and hardware Mitutoyo SJ400

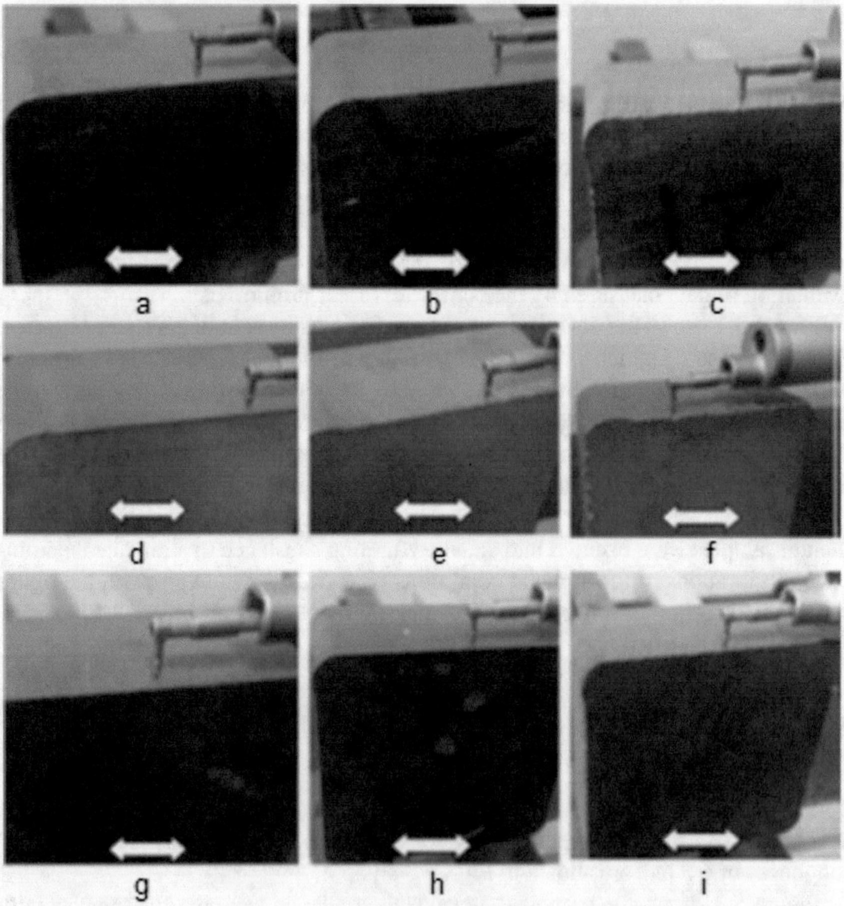

Fig. 5.2 Samples of stainless steel measured by contact roughness tester Mitutoyo SJ 400 **a**—AL120, **b**—AL220, **c**—AL370, **d**—SS050, **e**—SS120, **f**—SS150, **g**—FS050, **h**—FS100, **i**—FS150

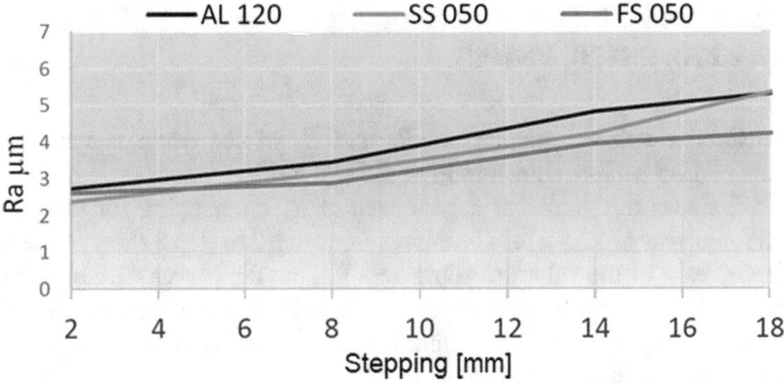

Fig. 5.3 Graph of R_{a} dependence on measurement stepping for samples with declared roughness of 6.3 µm

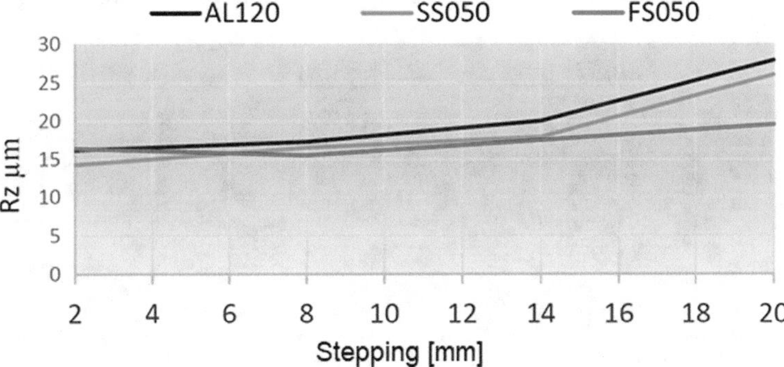

Fig. 5.4 Graph of R_{z} dependence on stepping of measurement for samples with declared roughness of 6.3 µm

SS050, a more moderate increase of roughness values can be detected because in dependence on material hardness the more increased roughness in this surface zone occurred.

Figure 5.4 shows the graphical dependence of R_{z} on stepping for samples with the declared roughness of 25 µm. The curves from step 2 to step 14 show a linear increase of values of roughness R_{z}. From step 14 the samples AL120 and SS050 show an increase in height values of irregularities in the last rough zone of the surface the cause of which is the cutting speed used for material hardness. In the case of sample FS050, a lower cutting speed was used and therefore the values of the R_{z} parameter are lower and linear development can be observed.

5.2 Samples with Declared Roughness of $R_a = 12.5\ \mu m$ (AL220, SS120, FS100)

The measured roughness parameters R_a and R_z of the samples with declared roughness of 12.5 μm are shown in Figs. 5.5 and 5.6.

Fig. 5.5 shows the graphical dependence of R_a on stepping for samples with declared roughness of 12.5 μm. The curves from step 2 to step 8 show in the smooth zone the increase of the values of roughness Ra, but the curves AL220 and FS100 document in the middle zone of surface (steps 8–14) a slight increase of the R_a values. In the case of the curve AL220, a slight increase of the R_a values can be observed from step 14 and in the case of sample SS120, the R_a values slightly decline. These changes represent the consequence of the use of a higher speed of cutting of samples.

Figure 5.6 shows the graphical dependence of R_z on stepping for the samples with declared roughness of 12.5 μm. The curves from step 2 to step 14 show a linear increase of values of roughness R_z in the smooth zone. In the medium rough and the

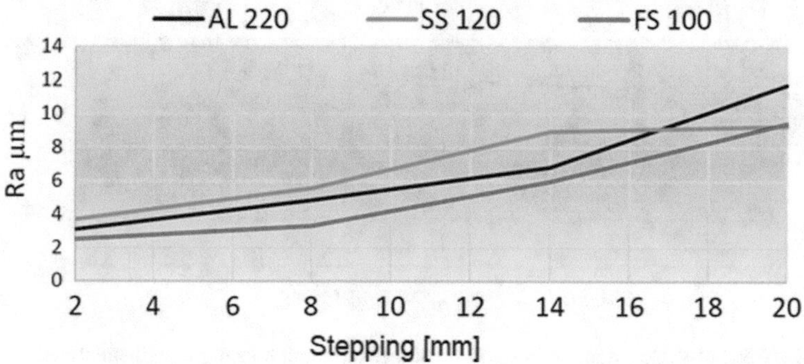

Fig. 5.5 Graph of dependence of R_a on stepping of measurement for samples with declared roughness of 12.5 μm

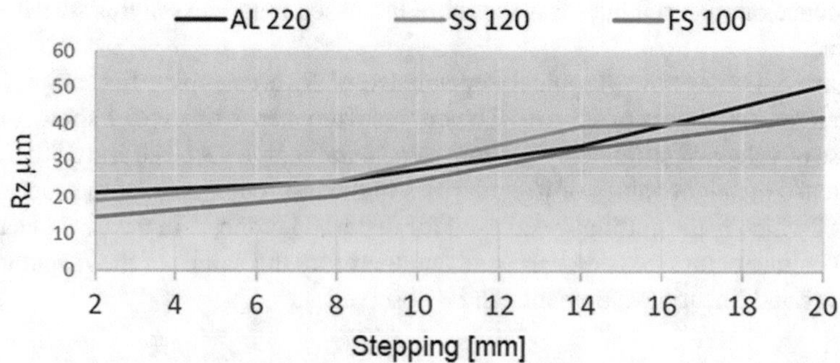

Fig. 5.6 Graph of dependence of R_z on stepping of measurement for samples with declared roughness of 12.5 μm

rough zone in the case of samples AL120 and FS050, the increase of values of R_z parameters can be observed. In the case of sample SS050 from step 14 decline or increase of the R_z values can be noticed which is probably caused by the crack due to the splitting of material off the sample surface.

5.3 Samples with Declared Roughness of $R_a = 25$ μm (AL370, SS150, FS150)

The measured roughness parameters R_a and R_z of the samples with declared roughness of 25 μm are shown in Figs. 5.7 and 5.8.

Figure 5.7 shows the graphical dependence of R_a on stepping for the samples with declared roughness of 25 μm. The curves from step 2 to step 8 in the smooth zone show a linear increase of the roughness values Ra. In the middle zone of the surface (steps 8–14) the curves SS150 and FS150 document a slight linear increase of the R_a

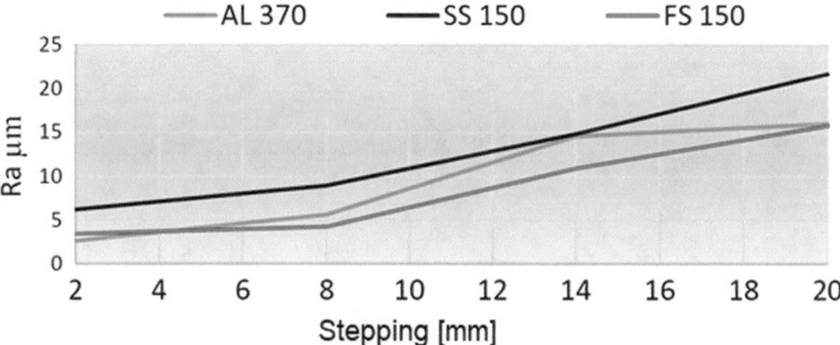

Fig. 5.7 Graph of dependence of R_a on stepping of measurement for the samples with declared roughness of 25 μm

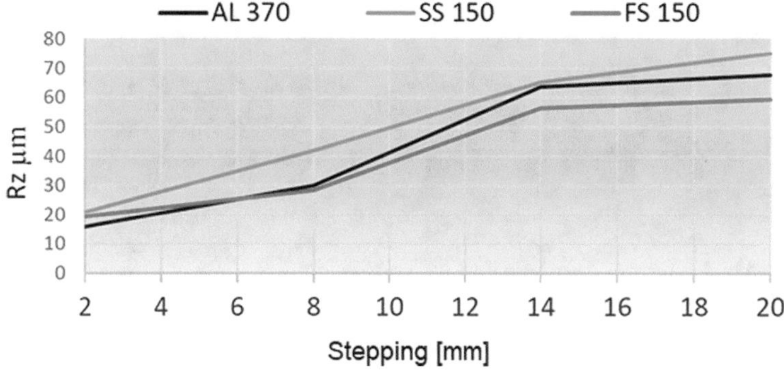

Fig. 5.8 Graph of R_z dependence on stepping of measurement for samples with declared roughness of 25 μm

values. In the last rough zone, the curve SS150 shows a slight increase and the curve FS150 shows a slight decline in the R_a values. The cause of the aforementioned fact rests in different hardness of materials. In the case of sample AL220 more remarkable decline in the values can be observed from step 14 which is caused by a crack in the last rough zone.

Figure 5.8 shows the graphical dependence of R_z on stepping for samples with declared roughness of 25 μm. In the smooth zone, the curves from step 2 to step 14 show a linear increase in values of the R_z roughness. In the case of samples, AL120 and FS050 from step 14 increase in values of the R_z parameter can be observed in medium rough and rough zones. In the case of sample SS050 from step, the decline or increase of the R_z values can be observed which is probably caused by a crack formed due to the splitting of the material of the sample surface.

Chapter 6
Measurement of Roughness Parameters by Contactless Laser Profilometer LPM

Using a contactless laser profilometer the surface of samples was measured in 190 steps with the step length of 0.11 μm on ten levels of the stepping interval with the step slide of 2 mm into the depth of surface material. The Gain mode (amplification of signal image in the camera) amounted to 1. Shutter time (exposition period determining the time during. which the surface is scanned) reached the value of 18.019 ms. After previous testing measurements of this surface type, these parameter setting values of the LPM equipment showed the clearest view of the image and the lowest image noise in the case of all materials of samples machined by the AWJ technology.

In case of measurement by the contactless LPM system, the experimental samples were inserted into an anti-vibrating substance (Fig. 6.1) to assure the influence of vibrations from the ambient environment and stability of the measured sample the machined surface of which imperfectly abutted against the working area of LPM and swinging of the sample on the working area of LPM during motion of stepping motors in measurement.

6.1 Samples with Declared Roughness of $R_a = 6.3$ μm (AL120, SS050, FS050)

Figure 6.2 shows 3D models of samples of material with declared roughness of R_a = 6.3 μm which were produced by putting together the series of measured profiles by the software of LPM view. The measured parameters of roughness R_a and R_z of samples with declared roughness are shown in Figs. 6.3 and 6.4.

Figure 6.3 shows graphical dependence R_a on stepping for samples with declared roughness of 6.3 μm. The curves in the smooth, medium smooth, and medium rough zone (steps 2–14) show a broader range of values of the R_a parameter caused by the reflection of laser light from the surface of samples with roughness lower than 5 μm.

J. Ružbarský, *Contactless System for Measurement and Evaluation of Machined Surfaces*, SpringerBriefs in Applied Sciences and Technology, https://doi.org/10.1007/978-3-031-08981-7_6

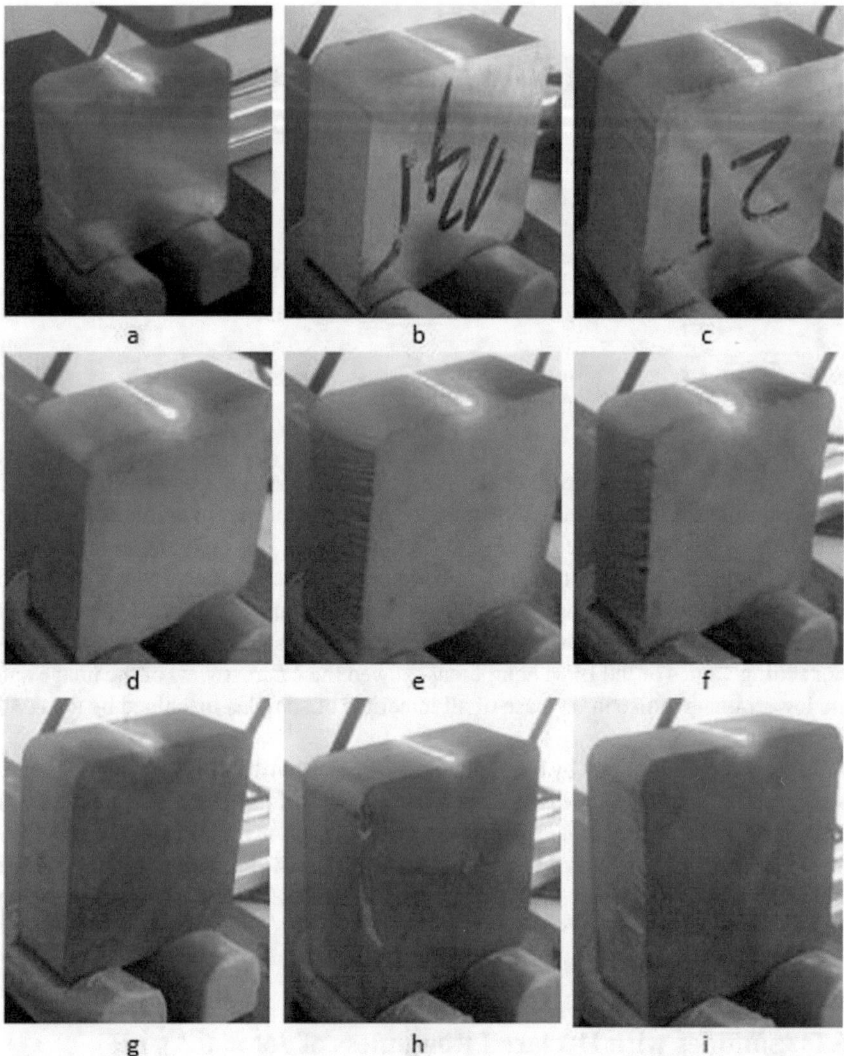

Fig. 6.1 Measurement of samples by contactless method produced by the AWJ technology with declared roughness according to marking. **a**—AL120, **b**—AL220, **c**—AL370, **d**—SS050, **e**—SS120, **f**—SS150, **g**—FS050, **h**—FS100, **i**—FS150

From step 14 the curves start to converge to a value of 6.5 μm which is caused by a weaker reflex directed to the CMOS camera.

Figure 6.4 shows the graphical dependence of R_z on stepping for the samples with declared roughness of 6.3 μm. Curves Al120 and FS050 in the medium rough and the rough zone show identical development. In the case of samples AL120 and SS050, the development is linear from step 14 which is caused by the dependence of setting parameters of AWJ cutting on material hardness. In the case of sample

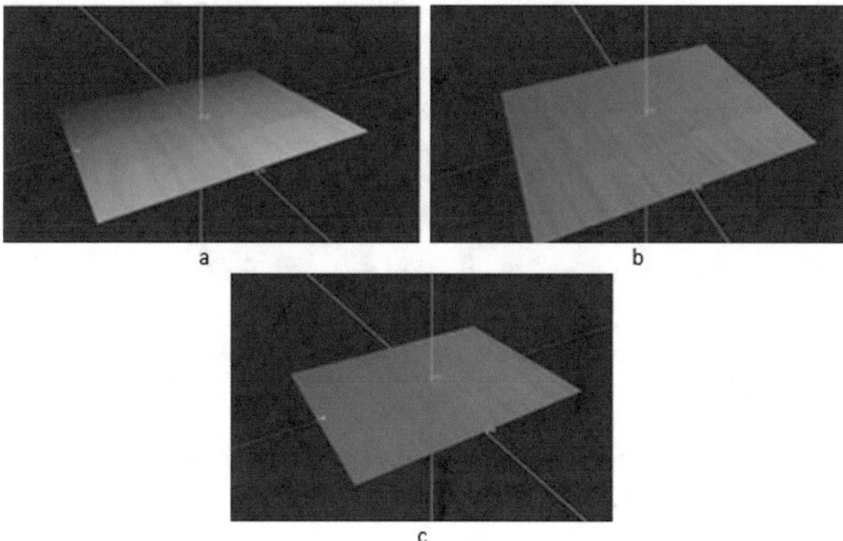

Fig. 6.2 3D model of the scanned samples of materials with declared roughness of Ra = 6.3 μm. **a**—AL 120, **b**—SS050, **c**—FS050

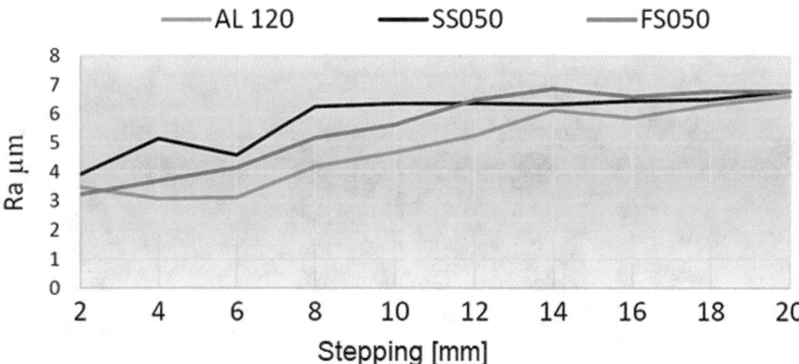

Fig. 6.3 Graph of dependence of R_a on stepping of measurement for samples with declared roughness of 6.3 μm

SS050, the increase of values can be explained as an undesired reflex of laser light to the camera.

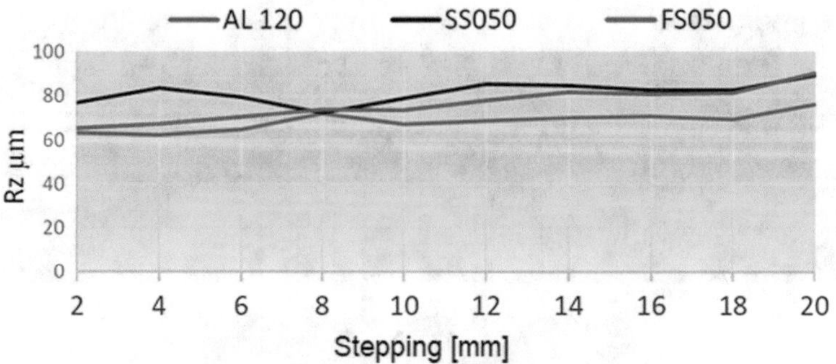

Fig. 6.4 Graph of R_z dependence on stepping of measurement for the samples with declared roughness of 6.3 μm

6.2 Samples with Declared Roughness of $R_a = 12.5$ μm (AL220, SS120, FS100)

Figure 6.5 shows the 3D models of material samples with declared roughness of $R_a = 12.5$ μm which were produced by putting together the series of measured profiles by the software of LPM view. The measured parameters of roughness R_a and R_z of samples with declared roughness of 12.5 μm are shown in Figs. 6.6 and 6.7.

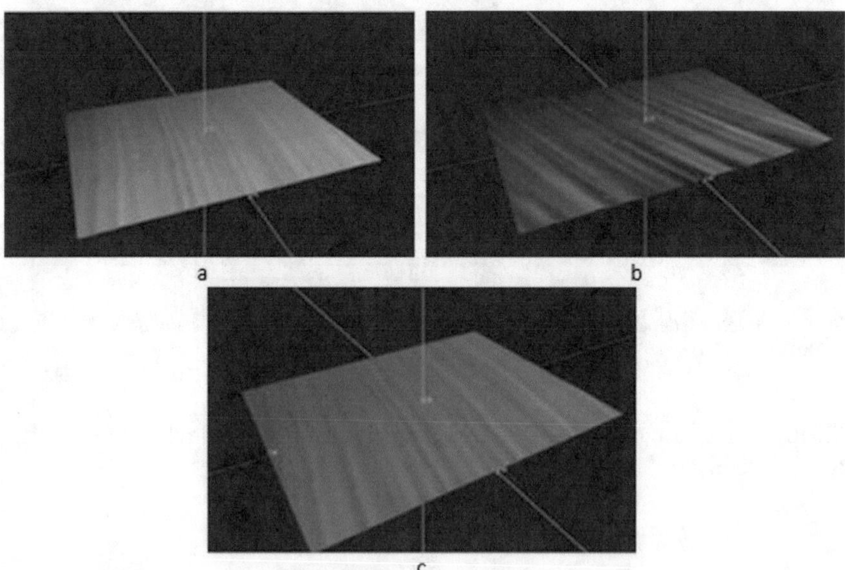

Fig. 6.5 3D model of the scanned samples of material with declared roughness of $R_a = 12.5$ μm. **a**—AL 220, **b**—SS120, **c**—FS100

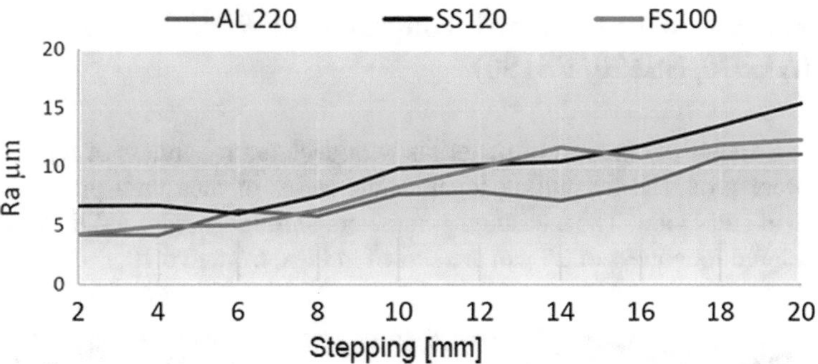

Fig. 6.6 Graph of R_a dependence on stepping of measurement for the samples with declared roughness of 12.5 μm

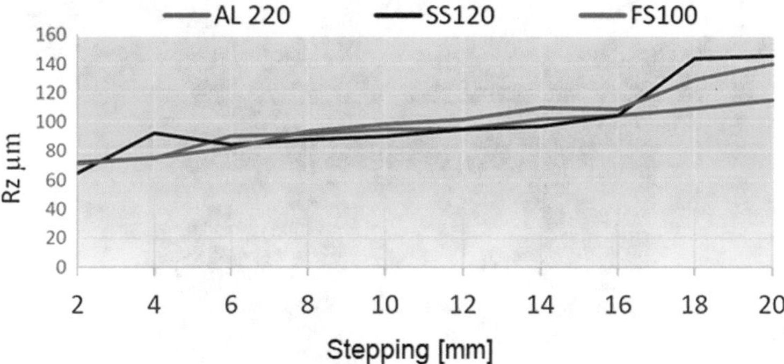

Fig. 6.7 Graph of R_z dependence on stepping of measurement for the samples with declared roughness of 12.5 μm

Figure 6.6 shows the graphical dependence of R_a on stepping for the samples with declared roughness of 12.5 μm. In the smooth zone curves, AL220 and FS100 show a linear increase of values of roughness R_a with negligible deviations. In the medium rough zone, the curve AL220 show lower values of R_a which are caused by the less intensive flow of water jet and abrasive through the sample, and in this zone the curve FS100 shows an increase of the R_a values caused by the more intensive flow of water jet and abrasive through this harder sample. The highest values of R_a are shown by the curve SS120 because the worst deformation could be observed in the channeled zone as this material proves to be the hardest of all experimental samples.

Figure 6.7 shows the graphical dependence of R_z of stepping for the samples with declared roughness of 12.5 μm. In the smooth zone, the curve SS120 shows the most visible variance. In the medium smooth and medium rough zone the curves AL220, SS120, and FS100 show linear character. In the smooth zone, the increase of the values in the case of the curves FS10 and SS120 could be observed which is caused by the fast removal of material resulting in higher values of parameters R_a and R_z.

6.3 Samples with Declared Roughness of $R_a = 25\ \mu m$ (AL370, SS150, FS150)

Figure 6.8 shows 3D models of materials with declared roughness of $R_a = 25\ \mu m$ which were produced by putting together the series of measured profiles by the software of LPM view. The measured parameters of roughness R_a and R_z of samples with declared roughness of 25 μm are shown in Figs. 6.9 and 6.10.

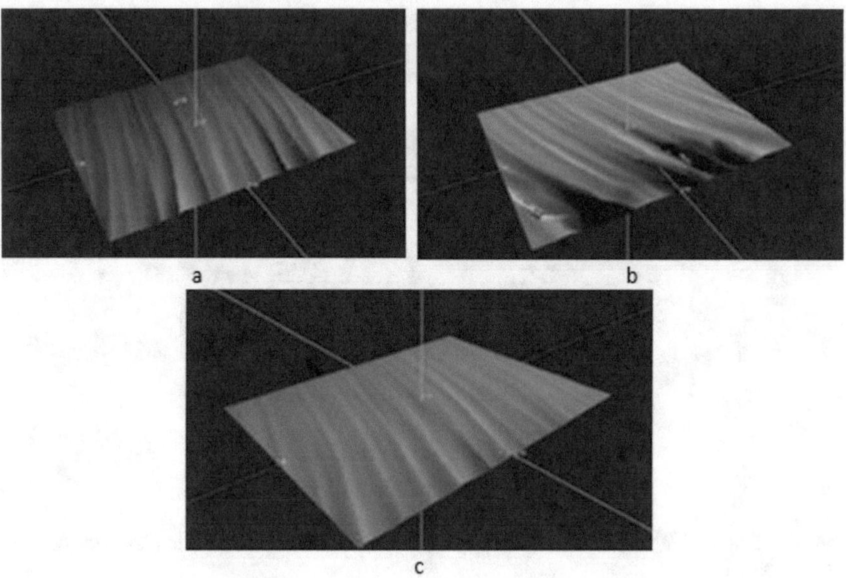

Fig. 6.8 3D model of the scanned samples of material with declared roughness of $R_a = 25\ \mu m$. **a**—AL 370, **b**—SS150, **c**—FS150

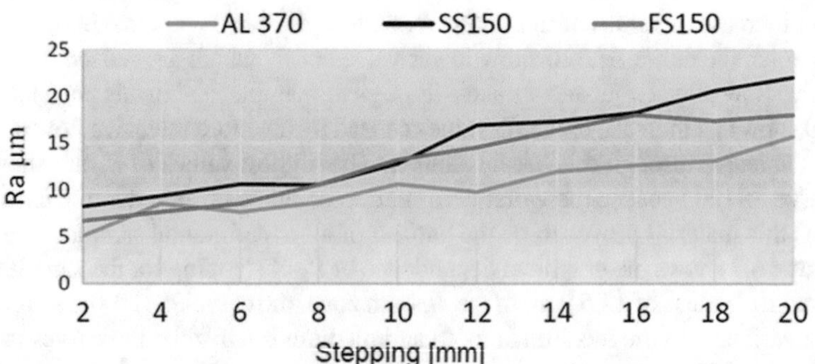

Fig. 6.9 Graph of R_a dependence on stepping of measurement for the samples with declared roughness of 25 μm

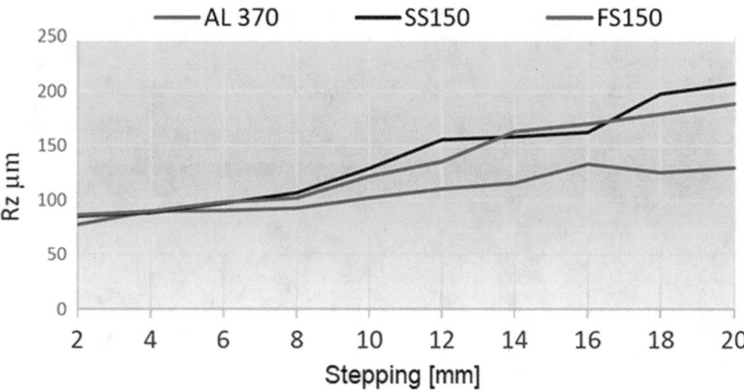

Fig. 6.10 Graph of R_z dependence on stepping of measurement for the samples with declared roughness of 25 μm

Figure 6.9 shows the graphical dependence of R_a on stepping for the samples with declared roughness of 25 μm. In smooth and medium smooth zone the curves dispose of the identical increase of values except for the curve AL370. The most considerable differences in values of the R_a parameter occurred in the medium rough and rough zone of the measured surface. It is caused by the higher speed of the cutting head of the machine during the production of samples resulting in a decline in the quality of the machined surface.

Figure 6.10 shows the graphical dependence of R_z on stepping for samples with declared roughness of 25 μm. The curves in the smooth zone show the R_z values increasing linearly. In medium smooth and medium rough zone the increase of values of curves SS150 and FS150 can be observed which reaches its peak in a rough zone where curve SS150 shows the highest measured values. This is proved by the fact that the cutting head speed of the machine must be adjusted to material hardness when achieving the required roughness.

The most considerable variance of values of roughness parameters R_a and R_z was detected in the medium rough and rough zone on the surface of samples with the highest speed of material cutting by the AWJ method which is proved by both contact and contactless measurement methods. It indicates the correctness of the values measured by the LPM system. The greatest disadvantage of the contact roughness tester Mitutoyo SJ 400 rests in the limited measurement of roughness to maximally 16 μm in case the system displayed measurement error caused by the considerable variance of roughness values.

The disadvantage of laser profilometry rests in complications in the case of measurement of samples with a glossy surface. Therefore our system was extended by blue laser which due to its advantages partially eliminates these problems. However, the problem was not eliminated therefore further experimental measurement of the sample surface with extreme glossiness was carried out.

Chapter 7
Glossy Surface Measurement

An extra part of the experiment was to determine the suitability of contactless laser profilometer LPM for the measurement of glossy surfaces. For glossy surface measurement, we produced the sample from stainless steel with the standard marking of A 304 (Fig. 7.1) using laser cutting technology in the company. Material A 304 was selected for disposing of the glossy surface during laser cutting which was the main requirement for the sample. The sample disposes of the same dimensions as samples produced by the AWJ technology, apart from thickness which is 5 mm, because this cutting technology is used in case of thinner material cutting with a thickness of up to 12 mm. Sample A 304 was measured by contact roughness tester Mitutoyo SJ 400 to determine the reference value of roughness parameter R_a and by contactless laser profilometer LPM. Based on these measurements graphical dependence of R_a on stepping of measurement was designed which shows differences of values of R_a parameter measured by both contact and contactless method.

When cutting the sample, the machine TruLaser 3040 by the company TRUMF (Fig. 7.2) was used. The parameters of the setting of the machine TruLaser 3040 during the cutting of sample A 304 are shown in Table 7.1.

In this part of the experiment, the roughness parameter R_a was evaluated which represents a sufficient datum for verification of the suitability of use of the contactless laser profilometer LPM for measurement of these types of surfaces. R_a evaluation was carried out in three zones of the surface, i.e. the smooth zone, the medium zone, and the rough zone. The measured surface was divided crosswise into three parts a, b, and c (Fig. 7.3). At first, the smooth (longitudinal, step 1) part was measured using three measurements in parts a, b, and c. Consequently, the medium (step 2) and the rough (step 3) surface parts were measured. These parts are divided into zones a, b, and c (Fig. 7.4).

Figure 7.5 shows in the window of the LPM software the measured profile of the glossy surface of sample A 304 produced in the aforementioned company using laser cutting technology. The sample surface is rather glossy which can be observed

J. Ružbarský, *Contactless System for Measurement and Evaluation of Machined Surfaces*, SpringerBriefs in Applied Sciences and Technology,
https://doi.org/10.1007/978-3-031-08981-7_7

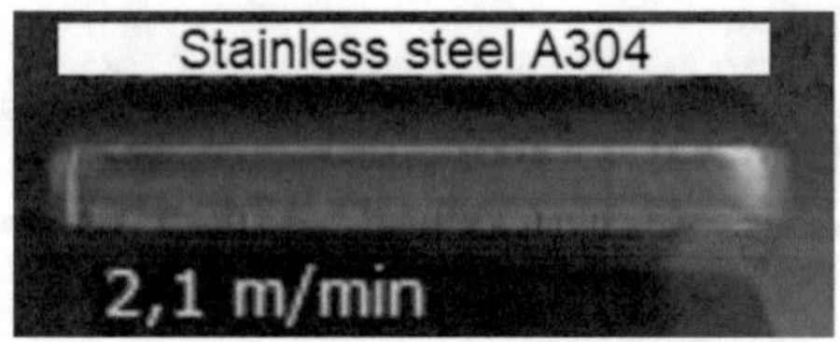

Fig. 7.1 Sample cut by laser in the company

Fig. 7.2 Machine TruLaser 3040 by the company TRUMPF designed for material cutting

Table 7.1 Parameters of machine TruLaser 3040 during cutting of sample A 304 by laser

Machine parameters	Stainless steel (A 304)
Laser performance	3200 W
Cutting speed	2.1 m/min
Max. cutting thickness	12 mm
Width of a cut – a gap	0.2 mm
Focus distance	4.5 mm
Diameter of focused ray, the diameter of rayon the nozzle	2.3 mm
Type and pressure of auxiliary gas	Nitrogen—17 Bar
Nozzle diameter	2.3 mm

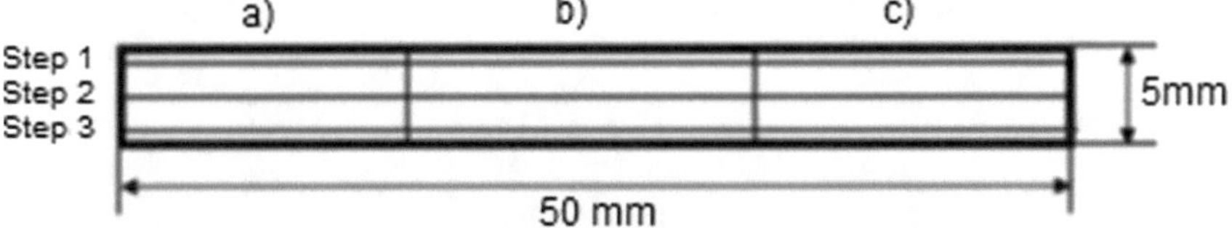

Fig. 7.3 Measurement of sample A 304 by contactless roughness tester Mitutoyo SJ 400

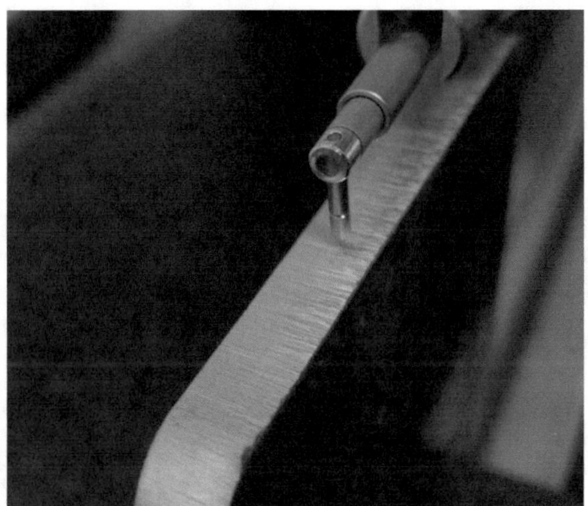

Fig. 7.4 Measurement of sample A 304 by contact roughness tester Mitutoyo SJ 400

Fig. 7.5 Profile curve of sample A 304 generated by the software LPM view

on the profile curve having bouncing development caused by the reflection of laser light from the measured surface back to the CMOS camera.

Figure 7.6 shows the graph of values measured both by contact and contactless method of sample A 304. The green curve represents reference values of R_a roughness measured by contact roughness tester Mitutoyo SJ 400. The red curve represents the values of R_a roughness measured by contactless laser profilometer LPM. The red curve proves lack of measurement by laser profilometry and surface glossiness caused reflection of laser from the sample surface to CMOS camera which resulted in a rapid increase of values of roughness parameter R_a, i.e. measurement error possible to be explained as undesired dazzling of the camera sensor. In the case of several measurements, this measurement error can be eliminated by deleting distorting values from the exported tables. In the case of this extremely glossy surface, the difficulty

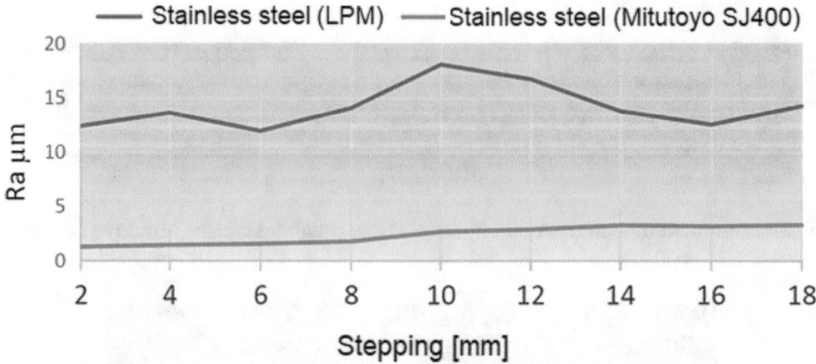

Fig. 7.6 Graph of values measured by contact and contactless method of samples produced by laser

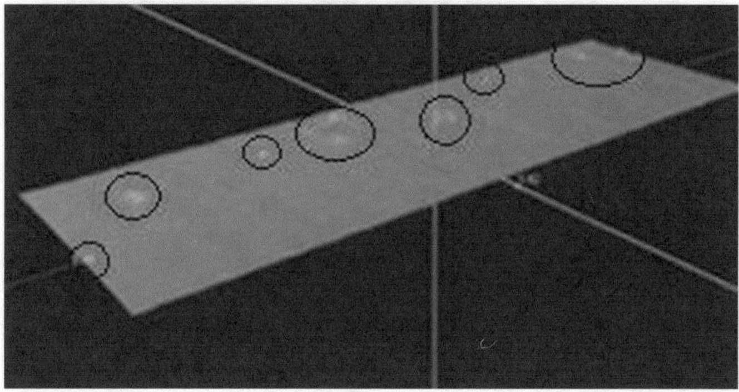

Fig. 7.7 Generated 3D profile of the sample A 304 produced by contactless laser profilometer LPM

cannot be solved as mentioned above, because the roughness highlighted in the red-colored curve exceeds the actual roughness value several times.

Figure 7.7 shows 3D models of the scanned samples made of aluminum and stainless steel produced using laser, which was designed by putting together the series of profiles measured by LPM view. The figures show the spots of the most noticeable reflection of laser light from the sample surface to the CMOS camera marked in a red circle.

These reflections were evaluated by the software LPM view as the largest irregularities of the surface and as the rapid increase of the R_a values.

7.1 Comparison of Achieved Results by Contact and Contactless Roughness Surface Testers

Graphical dependencies are divided according to the declared roughness of sample surfaces. Two graphical dependences are devised for declared roughness for parameters R_a and R_z. All dependencies contain six curves. Three curves for the values measured by contactless roughness tester LPM are connected into a single graph and compared with three curves of values measured by contact roughness tester Mitutoyo SJ 400. Comparison of experimental values measured by contact and contactless method consists of the creation of single-parametric dependences which represent the connection of values measured by contact and contactless method into independent graphical dependences for parameters of values R_a and R_z.

Figure 7.8 for the R_a parameter and Fig. 7.10 for the R_z parameter show comparison of values measured by contactless laser profilometer and contact roughness tester Mitutoyou SJ 400 for samples with declared roughness of 6.3 μm.

Figure 7.8 shows the graphical dependence of R_a on stepping of values measured by contact roughness tester Mitutoyo SJ 400 and by contactless laser profilometer LPM. The curves tend to slightly increase and the curves measured by the contactless method have higher values of the R_a parameter contrary to the curves measured by the contact method. It is assumed that it is caused by the reflection of laser light to the CMOS camera which was evaluated by the software LPM view as increased surface roughness R_a.

Figure 7.9 shows the graphical dependence of R_z on stepping of the values measured by contact roughness tester Mitutoyo SJ 400 and by contactless laser profilometer LPM. The curves develop linearly and the values measured by contactless methods are on average 215% higher than the values measured by the contact method. It is assumed that the values measured by the laser profilometer are increased due to the reflection of laser light to the CMOS camera which was evaluated by the software LPM view as the increased value R_z. The values measured by the contact roughness tester can be decreased because the diameter of the stylus of the measurement touch probe could not be inserted into the surface cracks with a smaller diameter. Despite the difference in values of the R_z parameter, these values dispose of identical increasing characters.

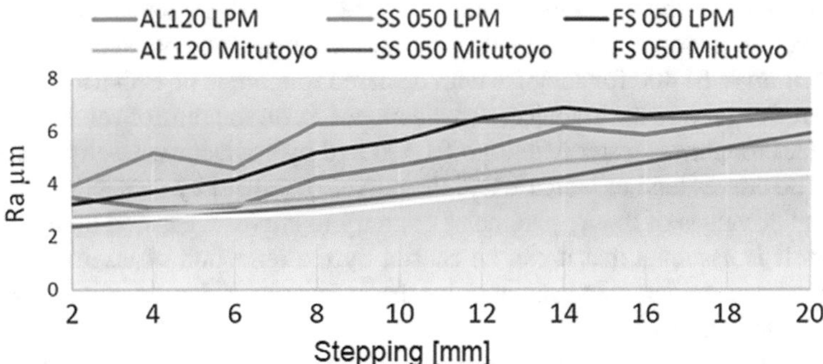

Fig. 7.8 Graph of R_a dependence on stepping of measurement of samples by contact and by contactless methods with declared roughness of 6.3 μm

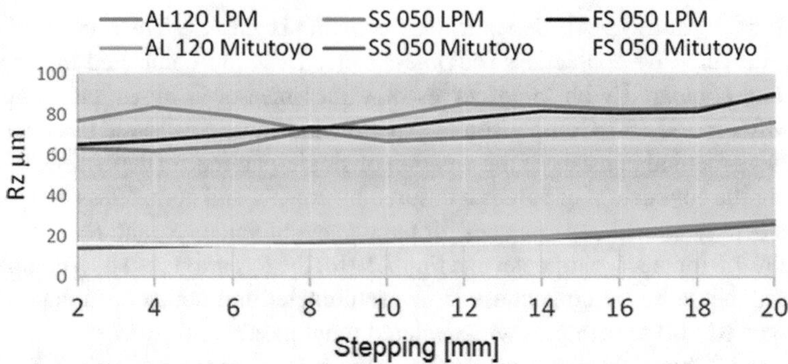

Fig. 7.9 Graph of dependences of R_z on stepping of measurement of samples measured by contact and contactless methods with declared roughness of 6.3 μm

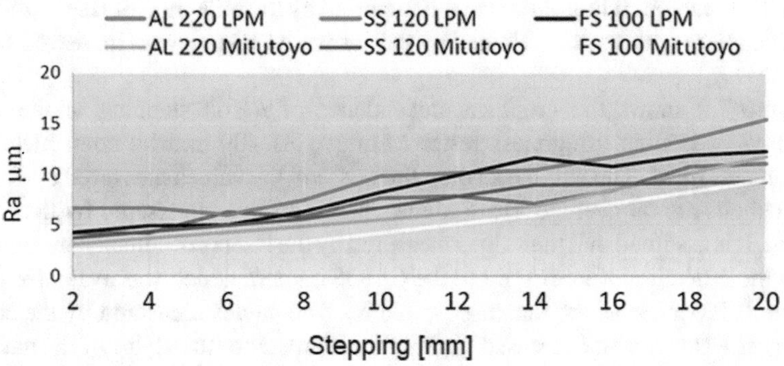

Fig. 7.10 Graph of dependence of R_a on stepping of measurement of samples measured by contact and by contactless methods with declared roughness of 12.5 μm

Figure 7.10 for the R_a parameter and Fig. 7.12 for the R_z parameter show comparison of values measured by contactless laser profilometer and by contact roughness tester Mitutoyo SJ 400 for samples with declared roughness of 12.5 μm.

Figure 7.10 shows the graphical dependence of R_a on stepping of values measured by contact roughness tester Mitutoyo SJ 400 and by contactless laser profilometer LPM. The curves tend to increase and the curves measured by contactless methods have higher values of the R_a parameter contrary to curves measured by the contact method. It is assumed that it can be caused by the reflection of laser light on the CMOS camera which was evaluated by the software LPM view as the increased surface roughness R_a.

Figure 7.11 shows the graphical dependence of R_z on stepping of the values measured by contact roughness tester Mitutoyo SJ 400 and by contactless laser profilometer LPM. The curves have an increasing tendency and the values measured by contactless methods are on average 200% higher than the values measured by contact method. It is assumed that the values measured by the laser profilometer

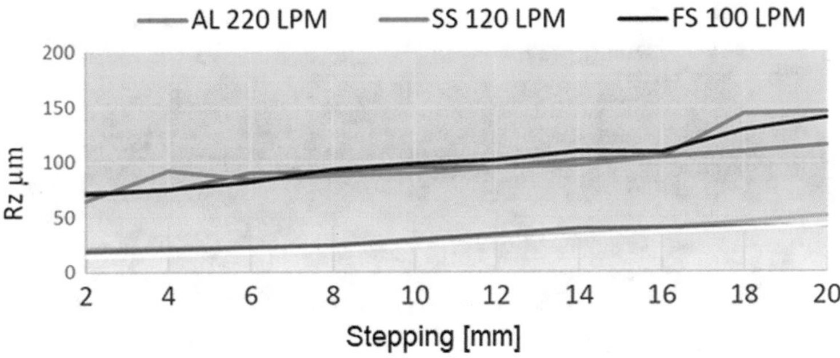

Fig. 7.11 Graph of dependence of R_z on stepping of measurement of samples measured by contact and by contactless methods with declared roughness of 12.5 μm

are increased due to the reflection of laser light to the CMOS camera which was evaluated by the software LPM view as the increased value R_z. The values measured by the contact roughness tester can be decreased because the diameter of the stylus of the measurement touch probe could not be inserted into the surface cracks with a smaller diameter. Despite the difference in values of the R_z parameter, these values dispose of identical increasing characters.

Figure 7.12 for the R_a parameter and Fig. 7.13 for the R_z parameter show comparison of values measured by contactless laser profilometer and by contact roughness tester Mitutoyo SJ 400 for samples with declared roughness of 25 μm.

Figure 7.12 shows the graphical dependence of R_a on stepping of values measured by contact roughness tester Mitutoyo SJ 400 and by contactless laser profilometer LPM. The curves tend to increase and the values measured by the contactless method show higher values of the R_a parameter than the values measured by the contact method. The curve AL370 in medium rough and in rough zone showed a decline

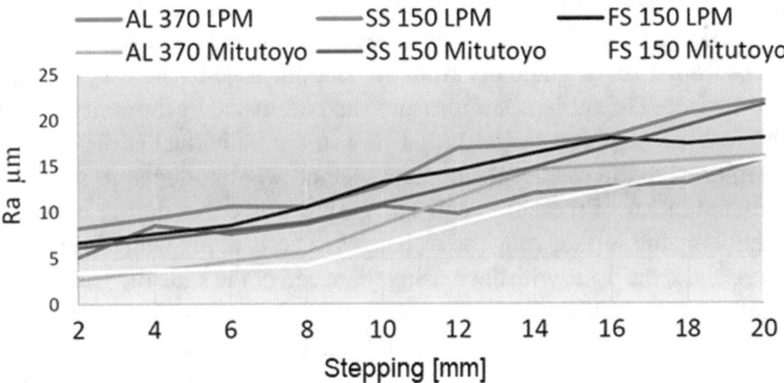

Fig. 7.12 Graph of dependence of R_a on stepping of measurement of samples measured by contact and by contactless methods with declared roughness of 25 μm

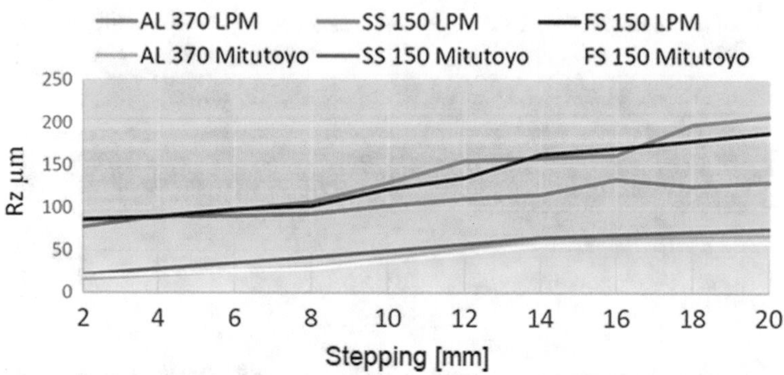

Fig. 7.13 Graph of dependence of R_z on stepping of measurement of samples measured by contact and by contactless methods with declared roughness of 25 μm

in values, which was caused by the formation of larger cracks due to higher cutting speed in this part of the sample surface.

Figure 7.13 shows the graphical dependence of R_z on stepping of the values measured by contact roughness tester Mitutoyo SJ 400 and by contactless laser profilometer LPM. The curves have an increasing tendency and the values measured by the contactless method are on average 170% higher than the values measured by the contact method. It is assumed that the values measured by the laser profilometer are increased due to the reflection of laser light to the CMOS camera which was evaluated by the software LPM view as the increased value R_z. The values of the R_z parameter measured by the contact roughness tester can be decreased because the diameter of the stylus of the measurement touch probe could not be inserted into the surface cracks with a smaller diameter.

The graphs of comparison of contactless and contact methods (Figs. 7.8, 7.9, 7.10, 7.11, 7.12 and 7.13) show the values of surface roughness R_a, R_z about stepping off the evaluated surface with the given speed of cutting head slide. The surface part measured and evaluated by contactless laser profilometer LPM fell into the range of stepping from 1 to 10 gradually from the smooth zone to the rough zone of the examined surface. The surface part measured and evaluated by the contact roughness tester fell into the stepping range from 1 to 4 in the individual surface zones. The stepping interval in the case of the contact method was shortened to steps 1–4 due to the demand factor of measurement. The graphs prove that during cutting by the AWJ technology the surface roughness changes linearly along with the cutting depth increase. At the same time, with decreasing the speed of the slide the area smooth and the medium smooth zone (characterized by lower values of roughness parameters R_a and R_z) gets enlarged which occurs especially in the first, the second, and the third cutting zone of the machined surface in case of samples with declared roughness of $R_a = 6.3$ μm (high-quality cutting). In the case of samples with declared roughness of $R_a = 12.5$ μm (medium cut), the area of smooth and of medium smooth zone occurs in the first and the second cutting zone of the machined surface and in the case with declared roughness of $R_a = 25$ μm (rough cut) the area of smooth zone occurs only in the first cutting zone of the machined surface. The values measured in the case of smooth zones show the remarkable difference in surface roughness independence on cutting head speed and material type.

Liu et al. [1] designed a similar contactless method of surface roughness measurement in the case in which they prove that the measuring process is to a certain degree influenced by the laser light source. Compared with the contact methods of roughness measurement even our contactless laser profilometer disposes of the disadvantage resting in trapping undesired reflections of laser light from the surface of glossy samples in the camera which are evaluated as increased surface roughness. In the case of samples AL120, AL220, SS050, SS120, FS050, and FS100 the values of parameters R_a and R_z increase linearly. In the case of samples, AL370, SS150, and FS150 the development of increasing values of parameters R_a and R_z are bouncing which is caused by the low quality of cutting due to the setting of a higher speed of cutting head during machining. Linearly increasing roughness can be thus included in a smooth, medium smooth, and medium rough zone which falls into the range of stepping from 1 to 7. The variance of values R_a and R_z can be observed in the medium rough zone within the range of stepping from 7 to 8. The highest variance of values R_a and R_z occurred in the last (rough) zone of the surface within the range of stepping from 9 to 10 in the case in which the water jet and abrasive tore off larger parts of the workpiece surface, especially in case of samples with declared roughness of $R_a = 25\ \mu m$ and that is a rough cut.

Following the comparison of cuttings and values of materials of samples, it is clear that in the case of material cutting technology AWJ the character of relief formation with soft (AL) and hard materials (SS) is the same. These results of measurements were recorded by both measurement methods. However, to achieve desired surface roughness with diverse physical properties of materials the cutting speed of the machine must be adjusted to surface roughness requirements. The surface roughness of samples produced by the company with the use of AWJ technology did not exceed declared roughness in the case of any of the employed samples. The biggest problem which caused a slight increase of R_a values and a higher increase of R_z values during measurement using a contactless laser profilometer were reflections of laser light occurring on the surface of samples which were recorded by the CMOS camera and consequently evaluated by the LPM software as increased surface roughness.

Following the comparison of graphical dependences for parameter R_a, it is clear that the values measured using the contactless laser profilometer are on average 2 μm higher than the values measured by contact roughness tester Mitutoyo SJ400. It is assumed that such a slight increase of the R_a values in the case of some steps is caused by the reflection of laser light from the sample surface to the CMOS camera. This measurement error can be partially eliminated by deleting distorting values from the exported tables. Complications occurring in the case of measurement of glossy surfaces by contactless laser profilometer were verified in another experimental measurement of sample A 304 produced using laser cutting technology. The sample surface was extremely glossy after being produced. The measured values were consequently compared with the values measured by the contact roughness tester Mitutoyo SJ400 and the values measured by the contactless laser profilometer ranged from 13 to 18 μm and the values measured by the contact roughness tester (considered the reference ones) ranged from 2 to 4 μm. According to this fact, it is clear that a contactless laser profilometer evaluates reflections as increased surface roughness, i.e. measurement error.

Following the comparison of graphical dependencies for the R_z parameter, it is clear that the values measured by the contactless laser profilometer are twice or even thrice higher than the values measured by the contact roughness tester. It is assumed that the reason for the increase rests in the reflection of laser light from the sample surface to the CMOS camera which was evaluated by the software LPM view as the increased value R_z. It is supposed that the R_z values measured by the contact roughness tester can be decreased because the diameter of the stylus of the measurement touch probe could not "enter" into each valley of the surface (the diameter of a tip of the measurement stylus is larger than the diameter of surface cracks). Despite the difference in measured values of the R_z parameter measured by contact and contactless methods, these values have the same increasing character.

Reference

1. Liu, J., Lu, E.H., Yi, H.A., Wang, M.H., Ao, P.: A new surface roughness measurement method based on a color distribution statistical matrix. Measurement **103**, 165–178 (2017). https://doi.org/10.1016/j.measurement.2017.02.036

Chapter 8
Conclusion

The publication deals with the research and development of a contactless system for measurement and evaluation of the selected feature of the machined surfaces. According to the existing structural solutions we designed and constructed the contactless laser profilometer intended for measurements and evaluation of surface roughness. The set-up of the contactless laser profilometer is divided into three parts as follows: mechanical, optical, and control parts. The mechanical part consists of a supporting frame structure made of satin anodized aluminum profiles. The optical part comprises the laser radiation source, the objective, and the camera with the CMOS sensor and the control part includes the PC with utility and evaluation software LPM view.

To determine the correctness of measurement by contactless laser profilometer LPM the samples made of three types of material (aluminium—EN 5083 standard, stainless steel—A304, and structural steel—S235) were subjected to the experiment of measurement of parameters R_a and R_z. The measured values of parameters R_a and R_z were processed and inserted into tables representing part of graphical dependences were generated as well which were compared with the values measured by contact roughness tester Mitutoyo SJ400. Following the comparison of graphical dependencies for the R_a parameter, it is clear that the values measured by the contactless laser profilometer are on average 2 μm higher than the values measured by contact roughness tester Mitutoyo SJ400. It is assumed that such a slight increase of the R_a values in the case of some steps is caused by the reflection of laser light from the sample surface to the CMOS camera. This measurement error can be partially eliminated by deleting distorting values from the exported tables. Complications occurring in the case of measurement of glossy surfaces by contactless laser profilometer were verified in another experimental measurement of sample A304 produced using laser cutting technology. The sample surface was extremely glossy after being produced. The measured values were consequently compared with the values measured by the contact roughness tester Mitutoyo SJ400 and the values measured by the contactless laser profilometer ranged from 13 to 18 μm and the values measured by the

J. Ružbarský, *Contactless System for Measurement and Evaluation of Machined Surfaces*, SpringerBriefs in Applied Sciences and Technology,
https://doi.org/10.1007/978-3-031-08981-7_8

contact roughness tester (considered to be the reference ones) ranged from 2 to 4 μm. According to this fact, it is clear that a contactless laser profilometer evaluates reflections as increased surface roughness, i.e. measurement error.

Following the comparison of graphical dependencies for the R_z parameter, it is clear that the values measured by the contactless laser profilometer are twice or even thrice higher than the values measured by the contact roughness tester. It is assumed that the reason for the increase rests in the reflection of laser light from the sample surface to the CMOS camera which was evaluated by the software LPM view as the increased value R_z. It is supposed that the R_z values measured by the contact roughness tester can be decreased because the diameter of the stylus of the measurement touch probe could not "enter" into each valley of the surface (the diameter of a tip of the measurement stylus is larger than the diameter of surface cracks). Despite the difference in measured values of the R_z parameter measured by contact and contactless methods, these values have the same increasing character.

The results of experimental measurement of the R_a parameter proved that the contactless laser profilometer measured the values which correspond with values of the R_a parameter measured by the contact roughness tester only with slight deviations caused by the undesired reflection of laser light from the measured sample surface to the CMOS camera. Consequently, the software LPM view evaluated these reflections as increased surface roughness. It is assumed that such a disadvantage could be solved by camera calibration or by the use of a laser light source with a different wavelength.

The values measured by contact and contactless method were compared with the declared roughness of the samples in case of which it was detected that the declared roughness was not exceeded by any of the samples.

The advantage of contactless laser profilometry rests in the repeatability of measurement, and the contactless and non-destructive character. Contrary to the contact method, the contactless method disposes of other advantages as well, e.g. the measurement tool, which after being worn measures incorrect data, does not get worn during the measuring. Another positive aspect is the working range of measurement allowing measurement of even more curved surfaces, unlike in the case of the contact method. The plus is also the fact that the LPM measures a certain area and then evaluates the roughness parameters whereas the contact method evaluates only part of the surface (straight line) along which the measurement touch probe moves. Vice versa, the disadvantage of the contactless method of measurement is a measurement of glossy surfaces, higher purchase price, measurement of roughness parameters is indirect and the measured values are interpreted in a more complicated fashion.

Realization of the solution of a contactless system intended for measurement and evaluation of surface roughness represents complementation and extension of measurement methods and evaluation of roughness parameters. The issue of contactless roughness measurement by laser profilometer is from the point of view of the method utilization for diverse types of material insufficiently explored therefore these queries shall represent an object of further research.

Uncited References

1. Raja, J., Muralikrishnan, B., Fu, S.: Recent advances in separation of roughness, waviness, and form. Precis. Eng.-J. Int. Soc. Precis. Eng. Nanotechnol. **26**(2), 222–235 (2002). https://doi.org/10.1016/S0141-6359(02)00103-4
2. Das, T., Bhattacharya, K.: Refractive index profilometry using the total internally reflected light field. Appl. Optics **56**(33), 9241–9246 (2017). https://doi.org/10.1364/AO.56.009241
3. Bass, M.: Handbook of Optics. Vol. I. Fundamentals, Techniques and Properties, 2nd edn. McGraw–Hill, New York (1995)
4. Glogowska, K., Majewski, Ł., Gajdoš, I., Mitaľ, G.: Assessment of the resistance to external factors of low-density polyethylene modified with natural fillers/Karolina Głogowska ... [et al.]—2017. Adv. Sci. Technol. Res. J. **11**(4), 35–40 (2017). ISSN 2299-8624
5. Mitaľ, G., Ružbarský, J.: Verification of the impact of finishing technology on the surface roughness. MM Sci. J. **2016**(October), 1100–1104 (2016). ISSN 1803-1269
6. Mitaľ, G., Ružbarský, J.: Diagnostics of selected surface characteristics with laser profilometry. MM Sci. J. **2018**(March), 2140–2143 (2018). ISSN 1805-0476. Spôsob prístupu: http://www.mmscience.eu/content/file/archives/MM_Science_201728.pdf
7. Holst, G.C., Lomheim, T.S.: CMOS/CCD senzore and kamera systéme. JCD Publishing, Winter Park (2007). ISBN 9780970774934
8. Bátora, B., Vasilko, K.: Obrobené povrchy. Technologická dedičnosť, funkčnosť. Trenčín 2000. ISBN 80-88914-19-1
9. Bátora, B., Vasilko, K.: Obrobené povrchy. Technologická dedičnosť, funkčnosť. Trenčín: Trenčianska univerzita, 183 s (2000). ISBN 80-88914-19-1
10. Balog, J., Chovanec, A., Kianicová, M.: Technická diagnostika. Trenčín: Trenčianka univerzita Alexandra Dubčeka v Trenčíne, 115s (2002). ISBN: 80-88914-66-3
11. Dovica, M. at al.: Metrológia v strojárstve. Košice: Technická univerzita v Košiciach, 349 s (2006). ISBN 80-8073-407-0
12. Hloch, S., Valíček, J.: Vplyv faktorov na topografiu povrchov vytvorených hydroabrazívnym delením. Prešov: Fakulta Výrobných technológií so sídlom v Prešove, 125 s (2008). ISBN 978-80-553-0091-7
13. Maňková, I.: Progresívne technológie. Košice: Vienala, 275 s (2000). ISBN 80-7099-430-4
14. Vasilko, K., Vasilková, D.: Metódy zlepšenia morfológie povrchov súčiastok. Prešov. Fakulta výrobných technológií so sídlom v Prešove, 120 s (2000). ISBN 80-7099-152-2
15. Vávra, P. at al.: Strojnícke tabuľky pre SPŠ Strojnícke. Bratislava,781 s (2009). ISBN 978-80-89223-28-2
16. http://cs.wikipedia.org/wiki/CCD
17. Szmicsková, M.: Drsnosť povrchu. [online]. Bratislava. Slovenský metrologický ústav. 2009. [26.1.2016]. Available on the Internet: http://www.smu.sk/storage/root/PDF/industryexpo2010/210/210_drsnost.pdf
18. Flowcorp.com: [online].Kent. USA. 2006–2010. Available on the Internet: http://www.flowcorp.com/waterjet-products.cfm?id=574
19. Daneshjo, N.: Klasifikácia technickej diagnostiky. TUKE. Letecká fakulta (2011). [online]. [20.3.2016]. Available on the Internet: http://www.sjf.tuke.sk/kpiam/taipvpp/2011/index.files/clanky/Naqib%20Daneshjo%20Andreas%20Kohla%20Christian%20Dietrich%20Mohamed%20Ali%20M%20Eldojali%20Klasifikacia.pdf
20. Fabian, S., Hloch, S.: Abrasive water jet process factors influence on stainless steel AISI 304 Macrogeometrical cutting duality. Scientific Bulletin, Volume XIX, North University of Baia Mare (2005). Romania. ISSN 1224-3264

Printed in the United States
by Baker & Taylor Publisher Services